SPARS AND RIGGING

FROM

NAUTICAL ROUTINE, 1849

JOHN McLEOD MURPHY
and
W. N. JEFFERS

Plate Drawings by
JOSEPH T. HIGGINS

"Learn of the little NAUTILUS to sail,
Spread the thin oar, and catch the driving gale.

* * * * * *

To instruct in what orbs the planets run,
Correct old Time and regulate the sun."—POPE.

DOVER PUBLICATIONS, INC.
Mineola, New York

Bibliographical Note

This Dover edition, first published in 2003, is a slightly corrected unabridged republication of the work originally published in 1933 by The Ship Model Society of Rhode Island, Providence under the title *Spars and Rigging: From Nautical Routine* which was, in turn, abstracted from *Nautical Routine and Stowage; with Short Rules in Navigation*, published in 1849 by Henry Spear, New York. The 1933 edition, of which this is the reprint, was limited to 200 copies.

The "Plates" section begins after page 90.

Library of Congress Cataloging-in-Publication Data

Murphy, John McLeod, 1827–1871.

Spars and rigging from Nautical routine, 1849 / John McLeod Murphy and W.N. Jeffers; plate drawings by Joseph T. Higgins.

p. cm.

Slightly corrected unabridged republication of the work originally published in 1933 ...under the title Spars and rigging.

Includes index.

ISBN 0-486-42989-X (pbk.)

1. Masts and rigging. I. Jeffers, William N. (William Nicholson), 1824–1883. II Murphy, John McLeod, 1827–1871. Nautical routine and stowage. III. Title.

VM531.M87 2003
623.8'62—dc21

2003046195

Manufactured in the United States of America
Dover Publications, Inc., 31 East 2nd Street, Mineola, N.Y. 11501

PREFACE.

THE aspiration of the man of patience, "Oh that mine enemy had written a book," was in itself harmless, compared with the wish, had it substituted the word *preface*—then, indeed, it would have been a sublimation of the patriarchal imprecation; and doubtless all who tread in the slippery and perilous pathway of authorship, can appreciate the pith of pious Job's revenge. Of late, however, in lieu of identifying these preliminary pages with the subject-matter of the volume, (to which, in fact, they should serve as indices,) it has become the fashion to embody in them a repetition of Paul Pry's apology of "I hope I don't intrude;" and, if one may speak from inference rather than experience, oftentimes the labor and difficulty of disguising this, under the form and garb of a passable *excuse* for publication, amounts to more than all the other toils of authorship. The exception to this may be found—*et exceptio probat regulum*—in the preface of an author who tells us that his object is to benefit the human species, which aforesaid information virtually implies a claim to the rank, style, and cognomen of philanthropist. Another, whose daguerreotype is exhibited in the *Charcoal Sketches*, under the title of the "*Best-Natured Man in the World*," modestly shuns the responsibility of individual independence, and as if realizing his loss of it for the first time, blames in his preface the *friends*, whose cries of "Publish it!" "Publish it!" he could not possibly resist.

For my own part, and the world may call me what they please, I am free to admit that a great portion of the matter contained in this work was gathered at sea, and while (to quote the bard of Stratford-on-Avon, who is strongly suspected of having had an eye to windward) engaged in that *Much Ado About Nothing*, incident to a cruise in "these piping times of peace." Subsequently these notes were arranged in form, with the view of preparing myself for examination; at the close of which, the dread alternative presented itself either of giving up the toils of my midshipman days to the cock-roaches, or that less book-loving fraternity, the *public;* and incited (like all other authors of course) by the hope, that as a printed manual of routine, they might be the means of imparting to the brows of my nautical brethren, a *wrinkle,* I hesitated no longer as to the course I should pursue. In view, then, of accomplishing this object, leave of absence was solicited from the Hon. Secretary of the Navy, and I set about the task of culling from a well-worn volume of half-obliterated notes, such as might be deemed worthy of publication. *Tupper* says, "Great is the dignity of authorship," and inasmuch as I confessed myself a convert to his "Philosophy," he might have spared me many a weary hour, had he added *great is the labor thereof,* for often the depth of my implication with the publisher proved a far greater incitement to "persevere unto the end," than all the promised "rewards" of a tenth edition.

In reference to my own labors, I am by no means insensible of the possible existence of imperfections in the order of their arrangement; but all who have clambered from the "kelson to the truck," will bear me out in the assertion, that the science of seamanship is characterized by one distinguishing feature, peculiar to itself—viz. that its guiding principles, in lieu of being established laws of observance, are rather matters of *opinion;* and though based upon the infallible criterions of reason and practice, as unsettled in their adaptation to case and circumstance, as the elements upon which they are born and cherished. This peculiarity, coupled with the fact that no man's professional career "has been so eventful as to have made him experienced in every thing," must necessarily beset with incalculable difficulties the path of one who attempts to delineate its outlines: and perhaps the whole science may be truthfully defined to be, the application of *common sense* to the equipment and management of ships. Add to this, dear reader, the additional fact, that though writing under the shelter of "a leave of absence," the vicissitudes of the service are ever such, that I could have no guarantee for an unmolested probation on shore, and naturally enough I was kept in constant trepidation, as it were, by a sword of Damocles, in shape of orders to sea—a circumstance not only calculated to exterminate the little ambition I felt to become an author, but to preclude, in some measure, that thorough treatment of the subject, which it was my design to give. But Pope says somewhere—

> "A hackney-coach may chance to spoil a thought,
> And then a nodding beam, or pig of lead,
> God knows, may hurt the very ablest head."

Nevertheless—to the point.

In the writing of the article on SPARS AND RIGGING, particular pains have been taken to render it as thorough as possible; and to refer to every improvement in marine equipments. For this reason, the best and most experienced riggers have been consulted, that their views might be obtained upon every point. Under these circumstances, I feel no hesitation in speaking boldly in its favor—for though emanating thus under my signature, it nevertheless embodies the opinions of men of the highest professional standing: knowledge of this fact, however, might have been gleaned from the absence of the *ipse dixit* throughout; for though presumptuous enough, as a Passed Midshipman, to write a book, I had neither the vanity nor the courage to associate the insignificant pronoun "*I*" with the detail of opinions and principles, of which many were in practical existence before "I" was born.

As an *ultimum vale,* I find it necessary to refer to some few comments, which I have made as notes, or otherwise, upon the works of other writers—more, however, with the view to remove any wrong impressions than to apologize for what I have considered my *duty*. With those gentlemen, I have not the honor of a personal acquaintance; neither have I regarded them as obstacles in my narrow and limited path of ambition, for I have no expectations or ends to serve in book-making, or authorship. What I have written has been done, *sans peur,* in full view and cognizance of the "golden rule."

JOHN McLEOD MURPHY, U. S. N.

THIRD-STREET, NEW YORK,
March, 1849.

INDEX.

SPARS AND RIGGING.

SPARS AND RIGGING.

SHEERS FOR LOWER-MASTS AND BOWSPRIT.

In the ordinary cases of masting, and particularly that of government vessels, the navy or dockyard sheers are resorted to, in lieu of temporary ones erected on board for the purpose. The manner of getting in, and stepping the masts by means of these standing sheers, is so simple, and so easily understood, that an explanation would in all probability be deemed superfluous. But as cases often occur, where it is necessary to erect sheers on board, the following method of rigging and raising them, is recommended, as that approved of by experienced and competent seamen.

Have a sufficiency of kentledge stowed to steady the ship for masting; shore the spar-deck fore and aft, best under the partner-beams, and sling fenders up and down the side, from the gunwale to the channels, and from the channels to the water-line; also, have skids from the gunwale to the deck, in-board, and chafing-mats on the quarter-galleries. Tow the sheer-legs alongside with their heads aft, and far enough to rest on the taffrail when on board. Lash a half-rounded spar on top of the gunwale, (well slushed on the upper surface,) for the parbuckles to lead over, when rove.

No. 1

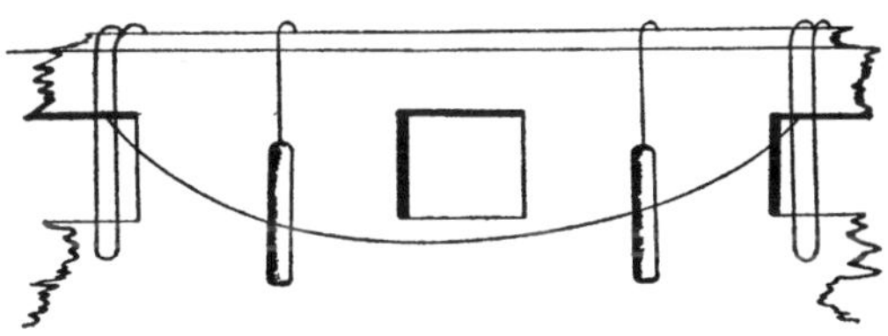

THE MAIN, OR OUTER PARBUCKLES,

Are usually formed from a stout hawser, which is middled, and the ends rove from *out, in,* a few ports apart, with the bight *outside.* (See No. 1.) Pass the ends over the gunwale, down, and under the sheer-leg, (one at the head, and the other at the heel;) thence up again, over the bulwarks, and snatch them to blocks in the opposite waterway, (abreast of their respective ports,) and lead the ends to the capstan.

No. 2

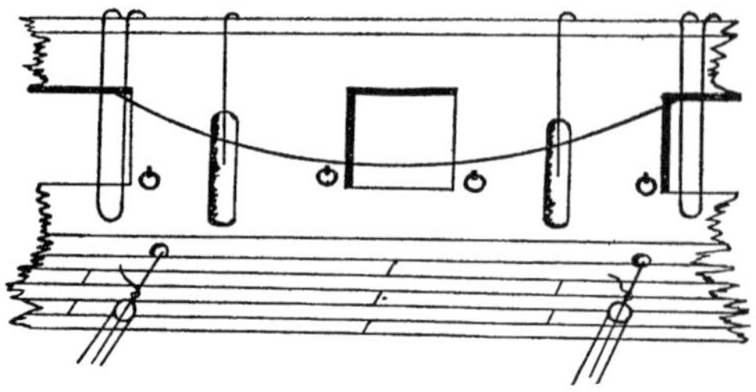

THE COUNTER, OR INNER PARBUCKLES,

Are rove through the same ports, from *in, out,* having the bights over the gunwale *in-board,* (see No. 2,) and the ends taken down through the gun-deck ports, (or scuppers, if a sloop-of-war,) immediately under those of the spar-deck, through which they are rove, and a round turn taken over spars lashed *fore and aft* in the ports, (for the purpose of easing the sheer-legs down,) having hands to attend them. Have a stout spar laid across the gunwale aft, to rest the heads of the sheer-legs on, when on board. Man the capstan; heave round and raise the leg up to the gunwale; reeve the counter parbuckles over it; haul them taut, and disengage it from the outer ones, lowering the heel on deck, and resting the head on the athwart-ship spar, abaft. The other is taken on board in a similar manner.

NOTE.—Lieut. Fordyce, of the Royal Navy, in his "Outlines of Naval Routine," in reference to the subject of sheers, says "When a vessel is confined to her own resources for spars to construct sheers, for getting in lower-masts, &c., the lower yards form the best—the yard-arms resting on deck should be fished, where they taper—woolded on, and the woolding set up by wedges. In a brig, the mainboom and mainyard are the best spars for sheers." It is almost needless to add, in corroboration of this idea, that the sheers for masting on board of merchant vessels are invariably constructed from the lower yards.

SHEERS.

Cross their heads, and square their heels, spreading them about two-thirds the breadth of beam at the mizen partners. Pass the sheer-head lashing (which should be of well-stretched rope) either on the bight around the cross of the sheer-legs, and expend the ends alternately, and knot them; or pass a few turns of the figure-of-eight first, and after expending the ends, either hitch them to the forks, or knot them as above. The opinions in reference to this point are so various, that it seems almost impossible which to designate as the best method, while many seamen advocate figure-of-eight turns altogether.*

Immediately under the cross, lash a large threefold block, for upper one of main purchase, and spread the heels to the waterways, to tauten the sheer-head lashing. The strap of this block is sometimes fitted with a long and short leg, (see No. 4;) the former is passed over the sheer-head lashing with a round turn—through the eye in the short leg—hitched around the fork of the sheer-head, and then seized

*The Board of Examiners of the late graduating class at the Naval School, were also in favor of figure-of-eight turn forms for sheer-head lashings.

with spun-yarn. Another way is to fit with two short eyes, which are secured under the cross by means of a lashing passed over that of the sheer-head. It should be remarked that these straps, however fitted, are *never* covered or served, in order that any indications of parting may be seen.

No. 3

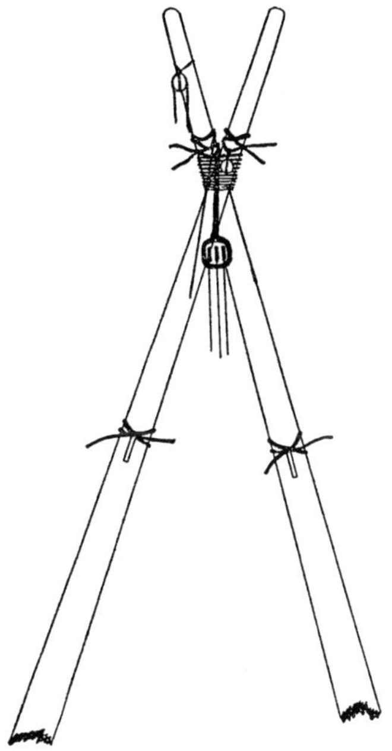

Next, lash a twofold block for upper one of small purchase to the after fork, having the cheek to rest against the sheer-head lashing, as a preventive from chafe. Higher up, on one of the forks, lash a girtline-block, (line rove,) for canting girtline: clove-hitch two good-sized hawsers around the cross, (see No. 3,) for forward and after head-guys; and about one-third the distance down each leg, another hawser for forward and after belly-guys, which should be cleated to prevent slipping. For SHOES, take two stout pieces of oak plank, of sufficient length to rest on three or four of the deck beams; cut mortices in them for the heels of the sheer legs to rest in, and have stout eye-bolts driven in them to lash the heels, and hook shoe-tackles. Place the heels in the mortices, and lash them to the eye-bolts; hook forward and after shoe and heel tackles to straps around the heels of the sheers; take the lower block of the main purchase, and toggle it in the bowsprit hole forward, or to the bitts, and reeve the fall by either of the following methods, which may be deemed preferable, viz.:

1. The end which you intend for the standing part, is rove through the side sheeve of the upper block, from *forward, aft,* through side of lower, middle sheeve of upper and lower, and so on, terminating at the side sheeve of the lower block; from thence it is hitched to the fork above the cross-lashing.

2. Reeve the end through the middle sheeve of the upper block, middle one of lower, side of upper and lower again; thence *across* to side sheeve of upper; and lastly, through the remaining sheeve of the lower block, hitching it, as in the former case, to the fork of the sheers.

Each system has its advocates, as well as its objections. In the first case, we avoid chafe at th risk of slueing the block in the strap; and in the second, *vice versa.* It is, however, best, in seamanship, as in morals, "of two evils to choose the least;" and to adopt the first method, since the turning of the block is only a *probability,* while the chafing of the falls, by reeving with a cross, is an inevitable *certainty.*

No. 4

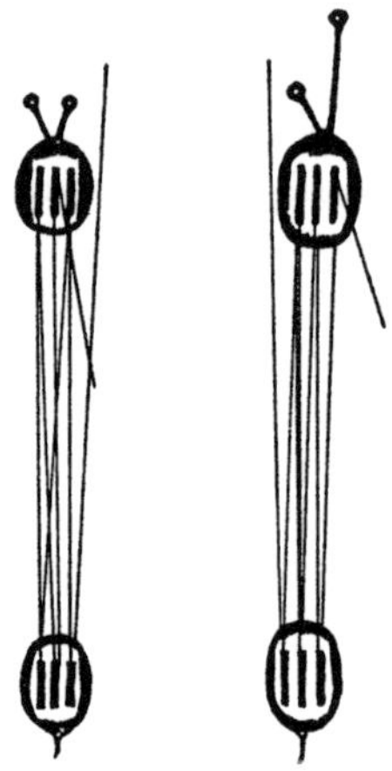

Having rove the purchase, pass the fall round the capstan; lead the guys forward, having a luff-upon-luff on each, and man them well. Launch the sheers as far aft as possible without tripping them; take aft the forward shoe and heel tackles, and set them well taut. Pass a stout lashing around the heels and over toggles, (*out side* of the ports;) also to ring and eye bolts near at hand. Secure them firmly, and heave away, catching the sheers as they rise with the athwartship spar. When nearly erect, avast heaving; lash the heels in the shoes securely, take the forward shoe and heel tackles forward, and man them, tending the after ones; cast off the bulwark lashings, and transport the sheers just forward of the mizen partners, having previously dampened the deck to facilitate the operation, (moving but one leg at a time,) and rake them, so that the main purchase will hang plumb over the forward part of the mast-hole.

Many systems have been devised by seamen as auxiliaries in raising sheers, but the following simple and effectual method is among the best.

No. 5

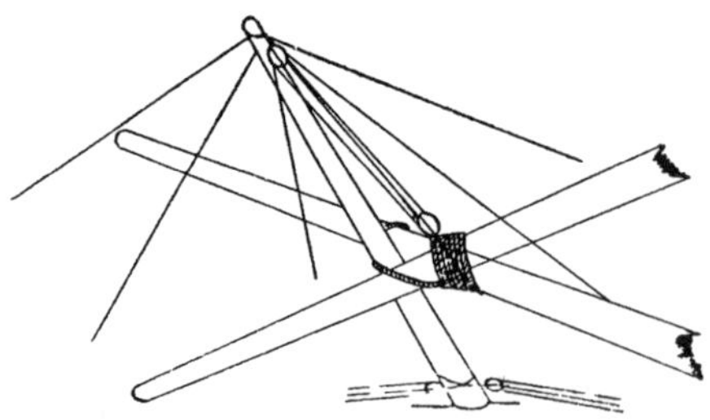

A small stout spar (say a stump top-gallant-mast) might be employed between the cross of the sheer-forks, where it could be retained by a loose lashing. Hook a stout tackle from the head of this spar to the sheers, and attach two other (canting and heel) jiggers together with head-guys, as in No. 5. With these, get it erect, slushing the spar and the forks at their points of contact. Now, with the assistance of the tackle, the head of the sheers can be elevated to a considerable degree, and the main purchase made to act, at an angle sufficiently great, to raise the sheers without further difficulty.

LOWER-MASTS.

The mizen is taken in first, to prevent the necessity of tautening the sheer-head lashing, which, on account of the decrease of beam, would become slack in transporting the sheers from forward, aft. Haul well taut the shoe and heel tackles; bring a strain on all the guys, and lash the heels to the bulwarks. Take the lower block of the main purchase through the sheer-legs *aft*, over the quarter, and toggle it to the garland on the mast, which should be towed alongside with the head aft.

As a diversity of opinion appears to prevail among seamen, as to the particular construction and positions of garlands, it is thought proper to add a few words in reference to this subject. Garlands, (represented in No. 6.) as employed on board of the French line-of-battle ship "*Hercule*," in taking out an injured mainmast, were formed of small stuff, (well stretched,) and passed around two posts, at a distance apart, and size, depending on the conditions of the case; the ends knotted, and the whole hitched over like a selvagee. The garland thus made, was placed on the *forward side* of the new mast *before launching*, at the distance of the *top of the gunwale*, to *the step in the kelson*, laid off *from* the *heel* towards the *head*. It was then secured to the mast by a cross-lashing as in the figure below, and the ends dogged, while a toggle was placed in the lower bight of the garland to prevent its rendering.

No. 6

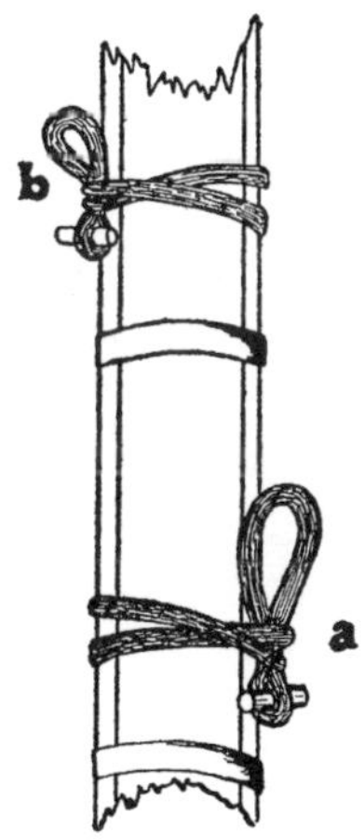

When both purchases are employed, as in the case above, in getting in heavy masts, a good plan, and one which obviates the necessity of heavers on the heel, is to lash the garlands a little on each side, and *not* in the same right line with the axis of the mast. Then, by slacking one purchase and holding on to the other, it may be slued at pleasure. The position of the small garland should be at the distance of *its purchase block, from that of the large one, on the sheers*, above the main, so that the falls cannot come two blocks except at the same time.

Man the capstan, (around which the purchases are passed,) and heave away, until the mast-head is up to the gunwale; lash a pair of single blocks to the tenon—reeve the girtlines—bend the canting girtline to the mast below the bibbs, and sway again until high enough; ease the heel in-board with a jigger or guy, coming up the after quarter, and belly-guys, (on the side on which the mast comes in,) which must be set up again. Bowse upon the canting girtline, and point the mast fair for stepping; wipe the heel dry; have carpenters below in attendance; lower away; white-lead (some old sailors say tar) the tenon and mortice, and step the mast. Pass a couple of straps around the mast-head—sway up pendant tackles—hook and set them taut, wedging the masts temporarily. Come up the purchases, off garlands, and proceed to transport the sheer-legs forward for taking in the main-mast. Cast off the bulwark lashings; man the guys and heel-tackles, and move one leg at a time. Rake and secure the sheers as before. Take the main and fore masts in, in the same manner as the mizen, with the additional use of the small purchase. If the sheers should be rather short, the forks may be spanned together by a piece of rope, passed from one to the other, and the upper block of the small purchase lashed to the span. Another way is, to rest the heel of the mast on stout blocks of wood, reaching across two or three of the deck beams, and after securing the head by means of guys, to shift the garlands.

The tops and tressle-trees, can be swayed on board with a burton on the sheer-head, and placed over before transporting, or afterwards with the girtlines and burtons.

In the merchant service, it is usual now-a-days, to get in the lower-masts before launching; and indeed frequently to rig the ship *in toto*. This, of course, saves much time, and (where the facilities are afforded) perhaps after all, answers the purpose quite as well. The manner of masting, however, in both branches of the service is essentially the same, and the leading principles as here stated, will apply to all the cases of ordinary practice. Under these circumstances, a particular reference to the differences and coincidences which occur, would, in all probability, be regarded as an idle and wearisome repetition.

BOWSPRIT.

Transport the sheer-legs forward, and rake the head over the bow, so that the lower block of the main purchase will hang just outside of the cutwater. Shift the upper block of the small purchase to the forward part of the sheers. Hook two stout tackles to the foremast head, and set them well up abaft; also two more from the foremast head to the sheers, or lash two large single blocks there instead, and reeve the after-guys through them, setting them up at the main bitts. Take the after-shoe and heel-tackles forward; tend the shoe-lashing, and lash afresh—setting all well taut. Secure the sheer-legs firmly to the bulwarks, and set the forward belly-guys up to the cathead. Previous to launching the bowsprit, put on the cap; also, the saddle for jib-boom, and lash the main-purchase garland on, so that it will hang *head* heavy, and the one for the small purchase just forward of the bees. Tow the bowsprit under the bow, and hook breasting-tackles from the eyebolts in the cap to each cathead, having a bedding-tackle in readiness also, to a strap around the heel.

The late Captain Glascock, in an excellent work, entitled the "*Naval Officers' Manual*," has added an *apropos* "caution" on taking in the bowsprit, which all the other *writers* (not riggers) appear to have omitted, viz.: "The main purchase must be overhauled down abaft the bumpkin; for should the purchase fall be passed before, and *under* the bumpkin, it will be impossible to lift the bowsprit *higher*. The spar must necessarily be lowered again into the water, the purchase untoggled, and shifted abaft the bumpkin, and, in fact, all the work must be done over again."

Man the purchases, and sway up until high enough for canting; then haul on the small purchase, hook the bedding-tackle, (which leads from the bitts through the bowsprit-hole,) point the heel fair, wipe it dry, white-lead the step, bowse on the breasting and bedding tackles, ease away handsomely the purchases, and step the bowsprit. Come up the purchases and tackles, and cast off the garlands; take the after-shoe and heel-tackles abaft; tend the forward ones; and mast-head tackles, (or hawsers;) cast off the bulwark lashings, and, by means of the after heel-tackles, rouse the heels of the sheers as far aft as the forward tackles will allow; then lower the sheers on the rail forward; strip them; hook the fore and main pendant-tackles to straps around the heads and heels; sway up and breast them over the gunwale with channel-jiggers, sending down the extra pendants (or hawsers) from the foremast head, and setting up the tackles as before, to steady the masts.

STANDING RIGGING.

Before a consideration of the subject of placing the rigging over the mast-heads, it is thought proper to refer in terms of brevity to the manner of cutting and fitting it, as being more directly in the natural order.

MAST PENDANTS,

When fitted, should be the length of their respective cross-trees; in which case, the *after* one would be the longer: but why this difference in length should subsist of necessity, is a matter now perhaps lost among the "dim traditions" of seamanship. Although the custom appears to have had its origin in the calamities incident to the sea; and the forward legs, by being shorter, were probably designed as a substitute, in the event of the yard-slings being shot (or carried) away; while the after ones, on account of their superior length, could be employed (in connection with tackles) to support the mast, without interfering with the yard, in case of a similar accident to the lower stays.

In fitting the pendants, turn a large thimble (well parcelled) in the ends of each, thrusting the strands of the splice once and a half; and after scraping them down, serve about a foot above; observing, in all cases, where a thimble is turned in, in this manner, to pass the turns of the service *towards* the thimble. Thus fitted, the pendants are wormed, parcelled, and served in the wake of the mast-head; after which, they are spanned together with small stuff on each side, enclosing half the square of the mast-head, from the centre of each.

The size of rope required for lower pendants is thus laid down in the tables, viz.:

Fore and *main*, 74's, 10½ inch; frigates, 10 to 9; sloops, 8 to 7½; brigs, schooners, &c., 6 inch.
Mizen, 74's, 8 inch; frigates, 7 to 6½; sloops, 6 to 5½.

NOTE.—In the merchant service, lower pendants are now seldom if ever met with; all heavy purchasing being done by means of triatics, and discharging-tackles. As this rule is only applicable to government vessels, it should be observed in reference to the pendants of small craft, (or where a single one is used, as on the mizenmast,) to fit them with a cut-splice, joining the two together on opposite sides of the mast-head.

LOWER-RIGGING.

The best and easiest rule to obtain the length of the shrouds, in the absence of a draft, is to take the mean distance from the top of the *opposite* lower tressle-tree across the mast-head to the partners of the mast: *plus, two-thirds* of the distance to the lower dead-eye. This will give the length of the first pair of shrouds, which should be laid off on the floor of the loft, and the shrouding (from which the rigging is to be cut) *flaked* closely around two samson posts, fixed at the distance of the extremities of the first *pair,* apart. In this manner, we obtain all the other shrouds in the order of their succession over the mast-head. Now mark each pair in the bight (forming the eyes) with marline, having a number of knots in it, for designation; and after cutting through the opposite end of the flake in a right line with the samson posts, get the rigging on a stretch.

From the centre of the eye down the leg, lay off a distance equal to the length of the after lower cross-tree, (that being the length of the futtock-rigging,) *plus,* one foot; at which, place a mark on each leg, and after worming, (with the lay,) put on the parcelling, beginning at the marks on the shroud, and working *up* to the crown of the eye, in both cases, in order to make the rigging shed the weather; then clap on the service, which is passed *against* the lay of the parcelling. At the distance of four feet from the end of the shroud-leg, with about one-half the diameter of the dead-eye in addition thereto, mark off, and turn in the dead-eyes *temporarily.* Take the *fifth square* of the mast-head for the circumference of the eye of the first pair of shrouds, and pass the EYE-SEIZING (Plate 1, Fig. 1), which is put on as follows: Heave both parts of the shroud closely together, by means of a strand, a little above the place where the seizing is to be put on, observing to clap on a layer of tarred parcelling, in order to protect the shrouding, and to make the turns lay fair and smooth. Splice an eye in one end, through which, and around both parts of the shroud, reeve the other, and pass the first turns (usually eight, being about the diameter of the rigging in width) *upwards towards* the crown; then thrust the end back, through the eye of the seizing, and the first and second turns, and pass the riding ones; after which, bring it down through the two last of the first turns, and frap all parts together, between the shroud-legs, forming a clove-hitch with the end in the frapping. All the other seizings are passed *from* the crown, once the diameter of the rigging *below* that passed on the previous pair, so that when the rigging is placed over, they may be situated in the same right line parallel to the tressle-tree. (Plate 1, Fig. 2).

The size of lower-rigging in merchant ships, depends as much on the proportion of spars as upon the class of vessels for which it is required; and although no universal rule can be laid down for the former, without the introduction of complex and lengthy tables, an assimilation *might* be made between the rules for government vessels and those destined for the merchant service. The size for the different classes is stated as follows, in the books, viz.:—

	74's	frigates	sloops-of-war	brigs, &c.
Fore,	10 inch.;	9 inch.;	8 to 7½;	6 inch.
Main,	" 10½ to 10	" 9	" 8 to 7½	" 6.
Mizen,	8	" 7 to 6½	" 6 to 5½.	

In the merchant service, it is usual to serve not only the forward swifters, but all the shrouds, the entire length. This custom, however, is not observed in the navy; a neat sword-mat being employed to protect the swifters.

N. B.—It should be remarked, that *all* the standing rigging used in the equipment of naval and mercantile vessels, is four-stranded right-hand rope; hawser-laid having gone out of date on account of its tendency to stretch.

MAIN AND MIZEN STAYS.

The rule as stated in the case of lower-rigging applies equally to the stays, with the additional length of the mast-head for the collars, the legs of which are laid up in the making, by leaving two of the strands long enough to double back and fit in the lay of the first half; while the bights thus formed by doubling, answer for the eyes. In fitting the collars, a backing of small yarns is used, and the *inside* ones from each strand are wormed into the stay some distance below the crotch; the legs are then wormed, parcelled, served, and leathered, down as far as (and to cover) the strand-worming on the stay.

The mizen, which is always single, is usually fitted with a shark's mouth for shackles on the mainmast, just above the deck; although frequently, in the merchant service, the lower end is *spliced* around the thimble of the shackle, and sets up by means of the eye-lashing aloft, Spanish-windlass fashion. In all other respects, the mizen is fitted like the main.

Again, the same difficulty is found in establishing a rule of size for merchant vessels, as in the case of LOWER-RIGGING; and in order to avoid lumbering up the narrow limits of the work with tables of reference, recourse must be had to the assimilation and comparison suggested in the preceding article, with due regard to the proportional disparity of masts. The rules for the naval service are thus laid down:—

	74's	frigates	sloops	brigs, &c.
Main,	15 to 14½ inch;	14 to 12½;	10 to 9½;	6½.
Mizen,	" 9½ to 9	" 9 to 8	" 7½ to 7¼.	

FORE-STAYS.

The rule as in the cutting, *size*, and fitting of main-stays, applies also to those on the foremast, with the exception of allowing *one-third* of the distance to where it sets up, in lieu of two, on account of the steeve of the bowsprit, and also the use of additional service, above the *nip*, to avoid chafe from the sheets and pendants forward.

Of late, a new system has been adopted in the mercantile marine, which, as yet has not found its way into the naval service. There, the lower-stays are of one entire piece, passing under the bowsprit, and cleated, instead of reeving through the bees or setting up on the ends (Plate 1, Fig. 5). They are fitted with lashing-eyes, and the crotch of the collar is formed by a seizing, binding the two stays together (Plate 1, Fig. 3). This method, whatever be its peculiar advantages—perhaps only those of economy,—is accompanied by the insuperable objection of bringing the entire weight and upward strain of the foremast, in heavy pitches, upon the gammoning and bobstay; besides obliging one to shorten and refit the legs, (in consequence of coming "eye and eye",) every time the stay requires setting up, which, as will be readily perceived, must be done aloft.

FORE-STAY STRAPS.

Although the method of reeving the stays through the bees has almost entirely superseded every other, the old custom of straps is sometimes adhered to; and as cases frequently occur where the bees split, or become otherwise injured, the following (but recently abandoned) is given, not only for the benefit of those whose prejudices for old systems still linger, but also in cases of accident to the stays when fitted after the "prevailing fashion." (Plate 1, Fig 4).

Take a piece of new rope, (eight-inch for a frigate,) stretch it well, and cut it in length *four* times the circumference of the bowsprit where it is intended to be placed, and *twice* that of the stay; get it on a stretch again, worm and cover it loosely with canvas; splice the ends together, middle it, and seize lock-thimbles large enough to receive the stay in one of the bights, having the splice to lay in the score. (Plate 2, Fig. 6).

In point of looks, these straps are somewhat disadvantageous when compared with the late method; and it should be borne in mind that the fore-stays, when reeving through the bees, require more length. and an additional allowance of the distance from the bolt in the bow, to the sheeve in the bees, must be made.

BOBSTAYS,

When of rope, are cut *twice* the length from their straps on the bowsprit, to the holes or shackles in the cutwater. They are wormed, parcelled, served, and leathered in the wake of the cutwater, and are fitted with a heart in the upper bight, which has the splice in the score, when rove. When chain is used instead of rope, they are only *once* the distance from the strap (or, in that case, *iron band*) to the shackle, *minus* one circumference of the bowsprit, which allows for a *rope laniard*, reeving through a heart, as in the method above. The heart is turned in with an eye-seizing, passed in the same manner as that referred to under the head of "LOWER-RIGGING." The size allowed for the vessels of the naval service is as follows (Plate 2, Fig. 7):—

74's, 10½ inch; frigates, 10 to 9; sloops, 8 to 7½; brigs, &c., 6.

BOWSPRIT SHROUDS.

It is best to fit them of rope and chain, (half of each,) and in length, once the distance of their eye-bolts in the bow, from the strap on the bowsprit. The rope part is fitted with a thimble in one end, which is connected to the chains by means of a shackle; the other extremity is spliced around a heart, and goes with a laniard, as in the case of bobstays. The splice of the thimble and heart should be served. In the merchant service, they are fitted entirely of chain, like the bobstays.

NOTE.—Care should be taken to exclude all service not absolutely essential to the preservation of the bowsprit rigging, on account of its retention of water, and the consequent tendency to produce rot.

FUTTOCK-RIGGING.

In length, equal to the half of lower cross-trees, fitted with a hook and thimble in one end, and an eye in the other—wormed, parcelled, and served the whole length, care being taken to keep them on a stretch as long as possible before setting up. *They are always the size of the topmast rigging.* Rope has, however, fallen into general disuse, and in nine cases out of ten, futtock-rigging is iron, except in the instance of large vessels, as ships-of-the-line, &c.

YARD-SLINGS

Are of chain, in length *twice* that of their respective mast-heads; to which must be added half the length of the forward lower cross-tree, that being the distance the yard should hang below the top.* The upper section, i. e. above the top, should be shackled together, keckled with a piece of large stuff to make it round, and neatly covered with canvas. The size of the chain necessarily depends on the weight it is required to support. (Plate 2, Fig. 10).

The yard-slings of merchant vessels are in length only half that of the forward lower cross-tree, and fitted with a hook which connects to an eye-bolt in an iron band going around the mast just below the eyes of the rigging (Plate 2, Fig. 9). By means of this plan, the yards are found to brace much easier than when suspended *vertically*, as done in the navy; while at the same time the slings acting at an angle serve to relieve in some measure the strain which is brought upon the truss by the sail bellying forward. Objections are, however, raised by many seamen, (notwithstanding the economy of the method,) on account of the necessity for slinging the yard *perpendicularly*, by a temporary lashing from the mast-head, in the event of having to house, or shift the topmast; since it is evident that the mast could not be stirred without resorting to this expedient, or to the more laborious one of sending down the yard to the deck beforehand.

TOPMAST SHROUDS

Should be cut the neat length from topmast-*head* to fid, and fitted in all respects like the lower-rigging, with the exception that the forward swifters are served all the way, (to avoid chafe from the topsail-yard,) and sister blocks for the lifts are

*It will be observed that the suggestion offered under the head of MAST-PENDANTS, in reference to their length, has a direct relation to this point.

seized between the first and second shrouds (Plate 2, Fig. 8; Plate 3, Fig. 14), just clear of the bolsters; and *not* between the swifter and first shroud, as recommended in some of the books; as the pressure of the yard, in clewing down, on the lee side is apt to split the shell; care being taken to seize the one on the *port* side the diameter of the rigging further *below,* so that it may be in the same right line with the one on the starboard side.

The size of rope for topmast rigging, varies much with the particular opinion the rigger may entertain with regard to its utility; for many seamen hold, that it is of little or no absolute service, except as a means of ascending and descending from aloft. But the contrary of this estimate is satisfactorily exemplified in the fact, that where vessels have been obliged to lose their topmasts, it is *always* necessary to cut away the laniards of the shrouds, after the backstays have been severed, in order to get rid of the mast. A reference to the following extract from the Tables of the naval service, compared with the size of lower-rigging previously stated, gives as a general rule an average of six-tenths.

Fore and main, 74's, 7 in.; frigates, 6 to 5½; sloops, 4½ to 4¼; brigs, &c., 4 to 3¾.
Mizen, " 5¼ " 5 to 4½ " 4¼ to 4.

TOPMAST AND SPRING STAYS.

For length, take the distance from the after-part of their respective mast-heads, to where *they lead* and set up (Plate 2, Fig. 11). They are fitted in all respects like the lower-stays, being leathered in the wake of the bees, clump-blocks, or whatever they reeve through. It should be remembered, however, that this rule has regard to the diversity of manner in leading stays; some being brought to the deck, (as in the case of those of the main-topmast,) while others are made to perform the service of bowsprit, shrouds, &c. And again, in the merchant vessels of the present day, they are sometimes fitted to go around the bowsprit, and cleated, like the lower-stays, already mentioned. The best manner of leading them, is to reeve the fore through the bees, and set them up to the bow—to bring *one* of the main to the deck, without regard to the foremast-head, and the mizen to the main-top.* The size we find laid down as follows, for the different classes of vessels:

Fore & main, 74's, 9½ in.; frigates, 9 to 8½; sloops, 7½ to 7; brigs, &c., 6¼ to 5½.
Mizen, " 6 " 5½ to 5 " 4¾ to 4½.

STANDING AND BREAST BACKSTAYS

Should be cut the neat length from the mast-head to their respective channels. The *standing* ones go over the mast-head like the shrouds, and are fitted in the same manner, being covered with raw-hide, or served in the wake of the tops and lower-yards, (when braced up.) The *breast-backstays* are single, fitted with a small neat eye, which is parcelled and marled over.† A good-sized double-block is turned in the lower end for the fall to reeve through; the bight forming the strap of the

*The mizen-topmast has no spring-stay, of course.
†The best plan; as it admits of their being sent down in port.

block, should be made long enough to turn it out in case of necessity without drawing the splice, having a seizing close down to the block.

The breast-backstays should be covered, or served in the same manner as the standing ones.

The Tables give us the following estimate sizes, viz:

Standing.	*Fore and main,*	74's, 10 inch; frigates, 9½ to 9; sloops, 8 to 7½; brigs, &c., 6½ to 4½.
	Mizen,	74's, 7½ inch; frig. 7¼ to 7; sloops, 6¾ to 6½.
Breast.	*Fore and main,*	" 8 to 7½; " 7; sloops, 7 to 6½; brigs, &c., 5½.

Merchant vessels are now seldom fitted wtih breast-backstays; the standing ones being made to perform the service of both. This plan, in addition to its economy, is not without reason; for it is manifest, that when sailing by the wind, the masts are strained forward in the direction of of the lee bow, and consequently require support from *abaft.* The size of the backstays is usually that of the lower rigging.

JIB-STAY.

Take the distance from the topmast-head, to where it leads and sets up, for the length. It is fitted either with a neat flemish eye in the upper end, and lashed to the collar of the topmast-head, or with legs, to go abaft the mast-head, as in the case of the other stays (Plate 3, Fig. 12). The latter method is preferable, however, for by lashing it to the topmast-stay, the entire weight and strain of *two* stays, one of which supporting a heavy sail, is brought upon *one* collar. In either case it is wormed and served, or leathered, a few feet below the crotch or lashing, to avoid chafe from the foot of the top-gallant-sail. The other end is pointed, and a small becket worked in, to bend a marrying-line. The size of the jib-stay for the different classes of vessels in the naval service, appears thus:

74's, 8 inch; frigates, 7½ to 7; sloops, 6 to 5¼; brigs, &c., 5.

TOP-GALLANT RIGGING.

The *shrouds* should be the neat length of their respective topmasts and top-gallant-masts, when set up in the top, (in which case futtocks are dispensed with;) otherwise, the length of the top-gallant-mast only; wormed, parcelled, and served in the wake of the cylinder, and below the futtock staff, except the forward one, which should be served the whole length. They are made to extend entirely across, and to fit so snugly as to require being driven on. The eyes are formed by means of seizings on each side—a bull's-eye being seized in between the first and second shrouds, just below, for the top-gallant lift. In the case of odd shrouds, (which are seldom met with in the merchant service,) they should be fitted with horse-shoe eyes or cut splices, or in one with the standing backstays. (Plate 3, Fig 15).

The size is stated as follows, viz.:—

Fore and main, 74's, 5 inch; frigates, 4½ to 4; sloops, 3½ to 3; brigs, &c., 2¾.
Mizen, " 3½ to 3¼ " 3 " 2¾.

The *standing backstays* are cut the neat length from the mast-head to the channels, and fitted with horse-shoe eyes, which are wormed, parcelled, and neatly covered with canvas (Plate 3, Fig. 13). They should also be leathered, or served in the wake of the tops and lower-yards, when braced up; and, in size, are somewhat larger than the shrouds, as will appear by the following extract from the Rigging Tables:—

Fore & main, 74's, 6 inch; frigates, 5½ to 5; sloops, 4½ to 4; brigs, &c., 3½ to 2¼.
Mizen, " 4½ " 4 to 3¾ " 3½ to 3¼.

The *breast-backstays* are the same length as the standing ones, and fitted with a small eye in the upper end, which is marled over; the other has a single block turned in it, for the fall to reeve through. They are leathered or served in the wake of the tops; and in size they should correspond with the top-gallant shrouds.

The *fore* and *aft stays* (in length) should be cut the mean distance from the mast-head, to where they lead and set up, of the same sized rope as their respective breast-backstays, except the flying-jib-stay, which is somewhat stouter. They are fitted with a neat round-eye, (in English vessels, sometimes a running eye,) to go snugly over the cylinder, which is wormed and covered with canvas,—needing no protection, on account of being put over *first*, except a few inches of parcelling and service, just below the splice of the eye.

The rule for the length, as stated above, is applicable to the various methods of leading the stays, as the stretching (if cut from the coil) allows for the eyes and ends. The *fore* reeves through the *outer* sheeves in the jib-boom end; the *main,* through a sheeve in the after chock of the fore-topmast cross-trees; and the *mizen,* through a hole or roller in the after part of the main lower-cap. The *flying-jib-stay* is cut and fitted in the same manner as the fore-top-gallant stay, with the addition of parcelling and service in the eye, (on account of going over *last,*) and leather for a few feet below, to avoid chafe from the foot of the royal. The lower end should be pointed, and a neat becket worked in to bend a marrying-line.

The size of the flying-jib stay is a trifle larger than the top-gallant shrouds, viz.:

for 74's, 5 inch; frigates, 4½ to 4¼; sloops, 4 to 3½; brigs, etc., 3½ to 3¼.

ROYAL RIGGING.

The *shrouds,* which are single, (Plate 3, Fig. 17), should be led into the top in lieu of setting up to the cross-trees, as they bring a heavy upward strain on the horns in lee-lurches, which might better be obviated, and more support given to the mast, by (lowering their point of security,) and having them to extend further below. In that case, cut them the neat length from the hounds of the royal-mast to the topmast fid; leather them in the wake of the jacks, cross-trees, and futtock-staves, and turn small single blocks in the ends for gun-tackle purchases, which will save the necessity of breast-backstays. (Plate 3, No. 16).

The *backstays* are the neat length from the hounds of the royal-mast to the deck, and are fitted in one with the shroud; the eye (which is wormed, parcelled, and covered with canvas) is formed by seizing the shroud and backstay together on each side, and made to set snugly over the hounds, a small bull's-eye or thimble being seized between the two for the royal lift. The *fore-and-aft stays* should be in length the neat distance to where they lead and set up, fitted in all respects like the top-gallant stays, being leathered in the wake of sheeves, rollers, &c. The size of royal rigging is thus laid down in the tables:—

Shrouds and Backstays.	*Fore and main,* For 74's, 4 inch; frigates 3½ to 3¼; sloops, 3 to 2½; brigs, &c., 2½.
Do. do.	*Mizen,* For 74's, 3 to 2¾; frig., 2½ to 2¼; sloops, 2¼ to 1¾.
Fore-and-aft Stays.	*Fore and main,* For 74's, 3¼ inch; frigates, 3 to 2¾; sloops, 2¾ to 2½; brigs, &c., 2¼ to 2.
Do. do.	*Mizen,* For 74's, 2¾; frigates, 2½ to 2¼; sloops, 2 to 1¾.

NOTE.—It will be borne in mind, that the foregoing measurements, in all cases, suppose the masts to be fidded.

In a review of the *approximate* rules just given, (for they claim to be little else,) it will appear that many of them are so general, and so widely framed, as to lack that dependence in accuracy which strict economy demands. But it must be remembered, that the prejudices and opinions of seamen in reference to rigging are so various and conflicting, that the system of each would require some particular and special rule of guidance. Under these circumstances, no *invariable one,* which will stand the infallible test of universal application, beyond a mere "rule of thumb," can be substituted for the absolute and proper measurement of drafts. Recourse ought, therefore, *always* to be had to the most accurate means; and no ship is complete in her outfit, which is permitted to sail without a draft of her spars and rigging.

The process of executing a draft, sufficiently accurate for practical purposes, consists simply in a reduction (by means of a scale) of the dimensions of the vessel's spars to a *side view,* showing the height and length; and also one of a *cross-section,* showing the circumference of mast-heads, &c., in order to illustrate the absolute spread and angular action of the shrouds and backstays.

The foregoing article may appear, in many respects, incomplete without a particular reference to other of the standing rigging than that comprised within its limits; but it has been found, that the exclusion of such as appertains to the yards, &c., here, will tend to obviate many of the wearisome repetitions, which, under a different arrangement, would be unavoidable hereafter. Sufficient, it is hoped, (however briefly and imperfectly expressed,) has been said, to illustrate such of the principles of equipment practised at present, both in the navy and mercantile marine, as was designed under this head, and to call the attention of the reader to such of the defects and sacrifices (suggested by false ideas of economy) as have been deemed necessary for his consideration. The objections offered here, it will be perceived, refer almost exclusively to the merchant service, and to that recklessness of appearances and security which is manifested to so great an extent, in carrying out some ruling passion of narrow-minded avarice, or false principle of frugality,—in all of which, their apology seems to consist, in lightness of spars and sails, and that disregard for symmetrical proportions, characteristic of vessels of war.

BOWSPRIT.

Within a few years a complete change has taken place in the rigging of vessels destined for the merchant service; and to such an extent has the introduction of iron been carried, that the system may be said to have been entirely revolutionized. The substitution of iron for rope in the fitting of yards and bowsprits particularly, has become so universal, that the rules of the two services are now quite distinct. The comparative use of iron in the navy, on the contrary, has been so slow and unfrequent, as to be even painful to observe; and in this respect, it is perhaps *far* behind the improved condition arrived at, in the economical equipment of merchant vessels. It is urged, however, that the "splinters," resulting from the use of iron, (in the event of action,) are so formidable, as to render an introduction of it, as universal as that observed in the merchant service, highly objectionable; and also, that the "give," or yielding properties, are so slight, when compared with the elasticity of rope, as to endanger the spars by the sudden check of heavy and violent lurches; while increased weight, and the influence of temperature, are arrayed in the same ranks of opposition. But the advocates of the "old gray" system seem to have lost sight of the idea of *rope laniards*, in connection with chain bobstays and bowsprit shrouds, and to be mindless of the many unanswerable refutations of practice and experiment in our commercial marine. Without digressing, however, from our subject, further than to point out such defects and omissions as are deemed most important, it is proposed to state the system of rigging in the navy, and to make reference under the proper heads, to the difference of equipment in merchant vessels, wherever such differences may be found to exist.

As the bowsprit is invariably rigged first, either for the purpose of getting it down in its place before gammoning, or otherwise it becomes necessary to prepare a stage under it for the men to stand on, which is rigged as follows: Secure the heads of two small spars (as top-gallant studding-sail booms, for example,) together in a lashing hung from the bowsprit end, and rest the heels on the head-rail, where they should be seized, to prevent their slipping, and lay boards across from one boom to the other. If the fore-stays should *not* reeve through the bees, as recommended in the preceding article, but be fitted with straps, then measure two-thirds the length of the bowsprit *from* the knight-heads, *out-board*, for the point to put the strap for the port-stay, fitted in the manner described on page 10, as done on board of the frigate "Cumberland," while flag-ship of the Mediterranean squadron (Plate 4, Fig. 18).

Place the strap on the bowsprit, with the bight *uppermost*, in order that it shall not turn, while setting up the stay, (which would be the case were the reverse of the rule observed,) and put the thimbles through the other two bights, bringing them up again, so that those for the fore-stay shall be on the port, and those for the spring-stay on the starboard side; being careful that the parts do not ride on the bowsprit, and to place a seizing close down to the bights. Next to this, put on the strap and *heart* for inner-bobstay; then strap the *heart* for each bowsprit shroud. Once the diameter of the bowsprit further out, place the strap and thimbles for the spring-stay, (fitted as in the other cases;) then strap and *heart* for outer-bobstay. Next, strap and thimbles, usually called "spectacles," for topmast and spring-stays to lead through—if used as outer bowsprit shrouds, and not rove in the bee-sheeves; otherwise a strap and heart for each of the latter. Lastly, just outside the bees, place the strap and heart for cap-bobstay.

Reeve, middle, and splice the bobstays, and seize in the hearts for a full due at the splice, (in the same manner as the eye-seizing passed on the shrouds of the lower-rigging,) having a strip of tarred canvas in the score, and stitching leather around in the wake of the cutwater. Splice the end of each bowsprit shroud around a heart, and reeve a few turns of the laniard (on the bight) to give the rigging a fit, and set up the bobstays and bowsprit shrouds *temporarily*. Strap the fore and foretop-bowline blocks, (the former to the fore-stay straps, the latter to eye-bolts in the bees,) and proceed to gammon the bowsprit.

NOTE.—In lieu of *straps* for bobstays and bowsprit shrouds, *iron bands* are in universal use on board of merchant vessels.

GAMMONING.

Although iron for gammoning has almost universally superseded rope, it is, nevertheless, liable to accident; and as a prejudice still exists in favor of the latter, among the "old time" sailors, the following method of passing it, is that most approved of by experienced riggers (Plate 4, Fig. 19):—

No. 7

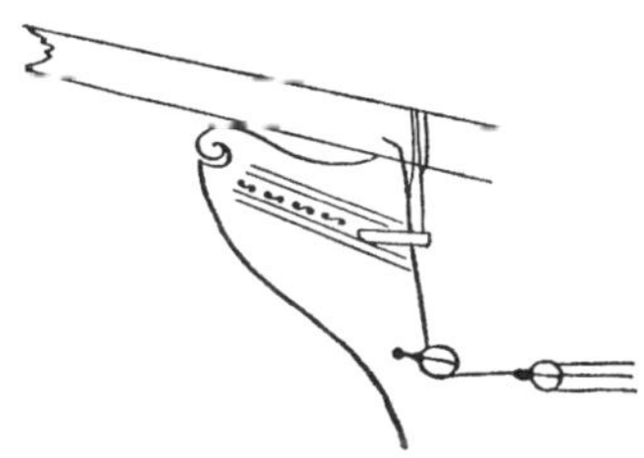

Tar and leather the bowsprit in the wake of the gammoning, having the latter laid on the upper surface, sufficiently wide to come back and cover all the turns when passed; and place the bolsters (usually of soft wood, well protected with layers of parcelling) in the *upper* part of the scuttle. Take a piece of well-stretched* *right-hand rope*, (which is preferable to hawser-laid, on account of the constant tendency of the latter to stretch,) and splice an eye in one end, and whip the other. Pass a turn of the gammoning around the bowsprit, *up* through the eye, over, and down, cleating it abaft to prevent its slipping, and set it well taut with a luff and pendant, which is fitted with a hook and thimble at each extremity, and rove through a single block secured to the bobstay-holes in the cutwater, by means of a strap sufficiently long, to allow the pendant to lead in a direct line through the hawse-hole, where a luff-tackle is clapped on, as illustrated in No. 7. Reeve a turn of the gammoning *against* the sun round the bowsprit and down

*"Or half-worn rope." —*Glascock.*

through the scuttle; form a bight with the end; thrust in a toggle, and put a strap around both parts above the latter, to which hook the outer end of the pendant, setting it well taut, and racking each turn respectively. Pass the balance of them, (usually seven,) in the same manner, from *aft, forward,* until the scuttle is filled; then take the end *back,* between the bowsprit and gammoning, and frap all parts together with it, using jiggers from each fore-tack bumpkin. Secure the end by a seizing; cut the rackings, (in order that each turn may share equally in the strain;) haul over the leather, (from *forward, aft,* that the inclination of the bowsprit may cause it to shed the spray,) and tack it down. If double gammoning is required, (as on board heavy ships,) pass the *inner* in the same manner—since, if the *outer* turns were passed last, the inner ones would become slack—and set up the bobstays and bowsprit shrouds for a *full due,* using a luff-upon-luff on *each* end of the laniard, as illustrated in the annexed drawing.

No. 8

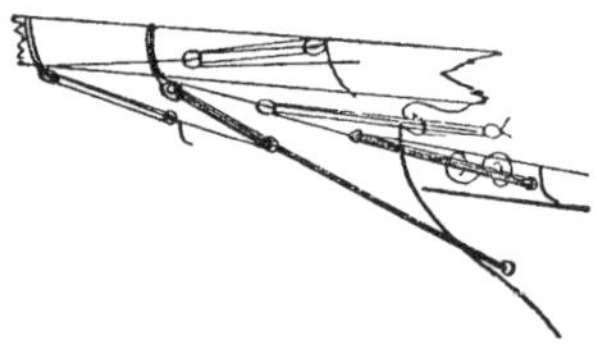

NOTE.—Of late, quite a number of merchant vessels have been fitted *without* gammoning, by giving additional strength to the step of the bowsprit. This, so far as *security* is concerned, has been found to answer all the conditions, and tends somewhat to vitiate the impressions formed with regard to the intensity of strain caused by the foremast. But in almost every case, there is a sacrifice of *length* in the bowsprit, (compared with the old *regime,*) which must contribute to affect materially the sailing capacities of a ship, by precluding that exposure of sail-surface forward, essential to rapidity of movement and ready obedience to the helm; besides imparting to a vessel a short and chubby appearance, far from pleasing to the eye of a seaman.

RIGGING LOWER-MASTS.

In getting up the LOWER TRESSLE-TREES—if not placed with the sheers, which is done by means of spans—remove the pendant-tackles mentioned on page 6, and send a couple of stout burtons to the mast-head, secured at each side, with the cheeks resting against the tenon.

Place the tressle-trees on deck, *forward* of the mast, and take out the *after*-chock, as the forward one, by having to support the heel of the topmast is more securely bolted, and consequently not designed to be removed. Hook the lower block of each burton to a strap around the *forward* horns, and stop the standing-parts along to the after ones; white-lead the mast-head and tressle-trees in the wake of each other; bend on a guy from forward, sway aloft, and as the horns rise above the bibbs, cut the stops, (a hand having been previously sent aloft,) and work them into their places. Send up the after-chock and bolt it. Now place and *lash* (not nail) the bolsters, having them made of soft wood, and covered with three or four thicknesses of tarred parcelling. Tar the mast-head in the wake of the rigging, and get over the lower pendants, which are swayed up by the burtons, hooking in the pendant-tackles, and setting them taut to steady the mast until rigged.

No. 9

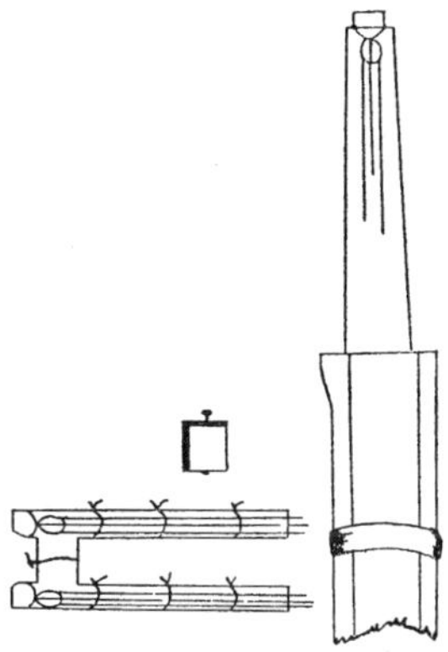

This plan is preferable to that of spans, on account of its security, and the perfect command which the burtons, by being stopped down to the tressle-trees, afford. In the merchant service they are invariably sent up by the sheers, almost simultaneously with the stepping of the mast.

WHOLE-TOPS.

Although half-tops have almost universally superseded the use of whole ones, except now and then in the merchant service, and the smaller classes of war vessels, the following plan of getting them over, is that adopted by an old and experienced rigger.

Have two stout girtlines secured at the tenon of the mast-head, and a good-sized single, or double tackle, hooked to a strap in the control part and directly between the girtlines, as in No. 10. Place the top on deck *abaft* the mast, with the forward part uppermost. Overhaul down the girtlines and tackle; pass the ends of the former underneath the rim and make them fast to their own parts, around the after-part of the top, stopping them out to each girtline-hole, as in the figure. Hook the lower block of the mast-head burton to a stout strap around the after-part, (to which a guy is also attached, *leading aft*,) and secure the *standing-parts* of the tackle and girtlines to the pigeon-hole by means of a squilgee-toggle, over which the bights are laid. Bend on a tripping-line to the toggle, (which should be greased,) man the burton and girtlines and sway away, pulling up steadily on all. When the forward rim comes up to the block, jerk on the tripping-line, (which disconnects the parts, and permits the girtlines to go to the side, and lead off fair;) sway on the tackle until the lubbers'-hole is clear of the mast-head, and lower away by means of the girtlines, sending it aft or forward with the tackle and guy as need be, having previously white-leaded the tressle-trees in the wake of the cross-trees, which are either secured to the top before sending it aloft, or sent up by means of the girtlines, first.

Bolt and secure the top with keys to the tressle-trees, ship the futtock and top-plates, stanchions, rail, etc., and send down the tackle and girtlines. It has been proposed to send two men up with the top, who hold the slip-ropes in their hands, and whose office is to cut the stop at the pigeon-hole; but this method was objected to by the late Board of Examiners at the Naval School, on the score of danger, and perhaps not without good and sufficient reason. The plan usually resorted to, is to send a man aloft beforehand, who cuts the stop; but even this, is liable to the same objection, and the first method, after all is said and done, is the best and safest.

No. 10*

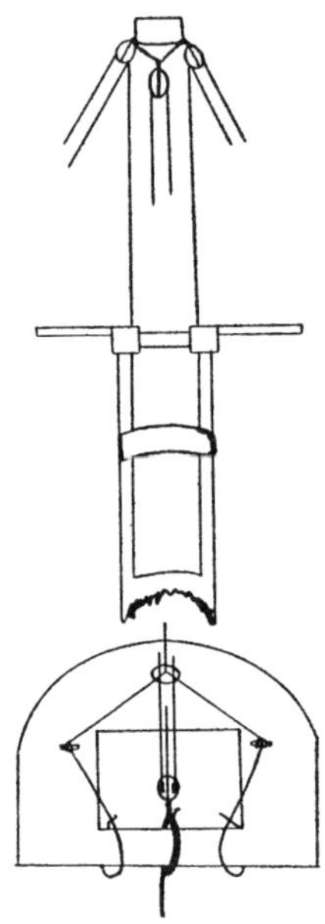

HALF-TOPS.

The following simple and beautiful method of sending up half-tops, suggested by a thorough seaman, (the late Boatswain Robert O'Neil of the navy,) is recommended as the best:

Have two stout burtons at the mast-head, (those used in getting up the tressle-trees,) and hook them to a strap through one of the futtock-holes in each half, which must be placed on deck, with the *outer* rims uppermost, at their respective sides of the mast. Sway the halves close up to the block at the mast-head and hang them there, until the cross-trees are sent aloft and bolted in their places. Then lower the halves down and secure them; sway up the upper cross-trees, and bolt and confine the whole with iron bands, sending down the burtons and girtlines, used to sway up the cross-trees. (See No. 11).

*It will be observed that the figure is somewhat at variance with the absolute conditions of the case, by having the girtlines and tackle apparently on the *upper*, instead of the *lower* side of the top; but this has been done designedly, that the reader might see the manner in which they were lead and secured.

No. 11

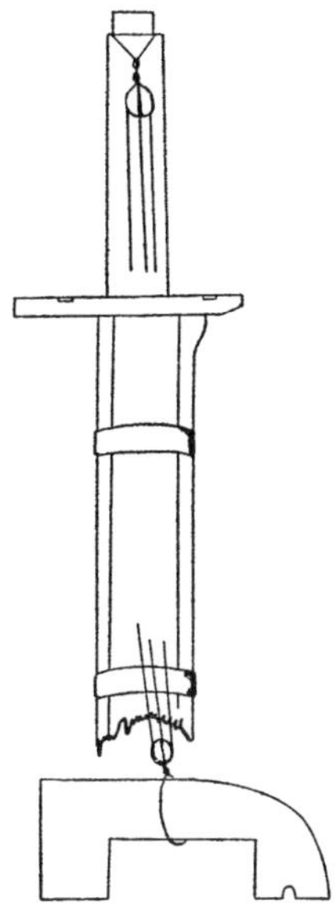

LOWER-RIGGING.

As the routine of rigging is nearly identical on all the masts, the method of one, e. g. the *fore*, will answer for a description of the others, care being taken to note the peculiarities of each as they occur. In nearly all the works on seamanship, it is recommended to shift the girtlines from the mast-head to the after-part of the tressle-trees, for the purpose of sending the shrouds aloft. But in the merchant service, where economy of time is regarded as a part and parcel of the profession, the topmast is pointed about four, or five feet above the lower mast-head, to which the eye-girtlines are attached, after the manner of a derrick. This plan is evidently preferable to that pursued by the riggers in the Navy-Yards, on account of the rapidity with which the shrouds can be got over, by saving the necessity of dipping and clearing the eye-girtline in each case. In heavy ships, however, two girtlines will be required to support the weight of the shroud, the block of the latter, or main girtline, being toggled to the *midship girtline-hole* in the top, as out of the way, and more convenient, than when secured to the after-part of the tressle-trees as directed in the books. Having done this, send hands aloft with marlinspikes, tar, slush, commander, etc. If the main-topmast and spring-stays are lead into the top, in lieu of leading directly to the deck, (as almost invariably done in the merchant service,) place the straps (Plate 5, Fig. 25) (fitted with clump-blocks, or thimbles and lashing-eyes, *to go forward*) over *first*,* in order to raise the eye-seizings of the under-shrouds, and to make them lay fair on the bolsters; the lower pendants having been placed, and *lashed* down, (so as not to disturb the fit of the rigging,) immediately after getting the tressle-trees aloft. Now proceed

*Brady says, "after the second pair of shrouds," but without assigning the reason.

to get the shrouds up, and over, in the order of their succession. *Knot* the ends of the main girtline together, and fit a toggle in one part, just above. Thrust this between the two parts of the first pair of starboard-forward-shrouds, from *out, in,* somewhat more than the length of the mast-head below the eye-seizing, and put a stop around both parts to retain the toggle in its place. Stop the girtline along the shroud towards the eye, and at the *crown*, and sway aloft. When as high as the top, bend on the small girtline just below the eye-seizing, taking the end from *in, out,* (if mast-head-girtlines are used, in lieu of one on the topmast as suggested,) and stop it as in the other cases. Cut the lower girtline adrift, as the shroud comes up, and steady it to the hand of the man aloft, who will gear the eye over the mast-head, and cast off the upper girtline. Place it fair on the bolster and beat it down with a wooden commander, being careful to carry the shroud well *aft*, as the angular action of the strain, in setting up, has a constant tendency to bring it forward. Turn the dead-eyes in *temporarily;* that is to say, by passing the throat-seizing only, (vide NOTE, p. 25,) and having the end stopped to the standing part; since if the quarter and end seizings were passed, the shrouds' liability to render in the throat-seizing would be greatly increased; and send the forward port pair up in the same manner. Now rouse the legs of the shrouds well down *amidships,* i. e. in a line parallel with the mast, to give the eyes a good fit upon the bolsters, and reeve a couple of turns of the laniards, (through both *side*-holes of the dead-eyes, in order that they may not slue,) and set up all four legs at the same time by means of the pendant-tackles, so as to keep the mast upright, and the eye-seizing fair.

All the other shrouds are got over in the manner described above, starboard and port alternately, and set up temporarily like the first two pair. Shift the girtlines from the top, up to the mast-head, and lash them to the sides and well aft. Dip them down through the lubbers-hole, and bend the starboard one to the fore-stay below the crotch, stopping it to the starboard leg; bend the port girtline on in the same manner to the port leg, and sway aloft, cutting the seizings as the legs reach the top. Pass a few turns of the lashing, (which is spliced into one of the eyes,) and either rest the collar on the lower-rigging, or hang it in a lashing from the mast-head. Send up the spring-stay in a similar manner, and dispose of it as in the other case. Now reeve the ends of the stays through the bee-sheeves, (or thimbles,) and prepare for setting up.

Hook a pendant-tackle to a couple of stout straps, one above the eyes of the rigging, and the other outside of the fore-stays on the bowsprit, with which the masts must be stayed as requisite. Fordyce has suggested that immediately succeeding the lower pendants, the *after-swifters* should be placed over, with the view of steadying the masts "gradually and accurately into their places," by means of them and the stays "without wringing the eyes or seizings at the mast-head." This idea is certainly a good one, and little perhaps is hazarded in the assertion, that had it been adopted more universally, many of the *outer* appearances in mast-heads, arising from the absence of some such counteracting effect, might have been obviated, and more than one rigger, competent in every other respect, justly saved from the censure of those who were unable to appreciate the absolute cause.

As there are considerations of magnitude involved in the subject of staying the masts, it is thought proper to pause for the introduction of a few "rough and ready" observations, which are offered for the reader's reflection.

Confining both the architect and the sailor to their respective spheres of action, the position, that is to say, the point in the interior of the vessel for fixing the step, is a problem, (in this *guessing* country of ours more frequently regulated by supposition than mathematics,) for the correct solution of which the former is alone responsible; while the latter, in his practical capacity at sea, becomes the instrument for testing the theoretical principles upon which the estimate is formed. Under these circumstances, designing only to treat of the subject of seamanshp, (in its peculiar and limited acceptation,) it would be foreign to the purpose of the work to attempt any inculcation beyond that which falls immediately within the province of the rigger.

1. Experiment has satisfactorily proved, that by raking the masts *forward*, we increase the vessel's tendency to pitch, by changing the direction of the effort exerted by the wind on the sails, from a line parallel to one forming an oblique angle with the keel. This, of course, should be guarded against, as offering an impediment to the progressive movements of the ship, *not*, however, so much in diminishing the force of acceleration, as in augmenting the spaces over which she must pass, by causing vertical deflections from a horizontal course; besides increasing the difficulty of trimming the yards, in consequence of their confinement when *by* the wind. Indeed, the sole advantage of inclining the masts forward, consists in producing a readiness to *wear*, which, however, is always attended with a corresponding indisposition to coming to, and the obvious necessity of lee helm, to make the vessel keep to the wind. In the ordinary evolutions at sea, these objectionable causes are not always self-evident, in consequence of producing effects which are correspondingly imperceptible. But in cases of emergency; e. g. as in scudding, (where the remedy is precluded,) they are often fearfully realized; and by preserving the disposition to fall off, we increase a thousand-fold the liabilities of being brought by the lee.

2 When the masts are stayed in a *perpendicular* position with respect to the keel, the effort exerted by the wind on the sails, will be in a horizontal direction. Under these circumstances, her progressive movements will be uniform, and the resistance and velocity constant; while the objectionable features visible in the face of the preceding system, are entirely obviated, and every favorable effort communicated, which can ensure speed and safety.

Lastly; where the masts are made to rake *aft*, an increase of after-sail is thus given to the ship; as also a disposition to approach (rather than recede from) the wind; by which we obtain rapidity of movement, in the evolution of tacking. The effects produced being nearly conversely to those which are developed by raking the masts in an opposite direction, and more favorable to the sailing capacities of the ship, as we obviate to a great extent the tendency to pitch, and the difficulties of bracing.

Having stayed the masts by means of the tackle referred to, (and the swifters, if the methods be adopted,) wedge them, and put on the MAST-COATS, which (according to *Glascock's* rule) are cut and made as follows, the material being either No. 1 or No. 2 canvas: "Girth the mast round at eighteen inches above the deck. Girth also round the deck at three inches from the mast-hole. These girths give the circumference at top and bottom. The length is obtained by measuring straight the distance between the places girthed. Divide the lower girths into an equal number of parts, suitably to the width of the canvas, allowing for the *seams*, which are one inch wide. The cloths must be gored upward, to produce the circumference of the mast at the top girth; and, when sewed together, cut with a sweep to set neatly round the mast. The upper part is then sewed into a double collar six inches wide. To find the quantity of canvas in a mast-coat, multiply the number of cloths by the length, and add the quantity in the collar."

EXAMPLE

	Ft.	In.	
	1	8	length.
		4	number of cloths.
	6	8	
	3	0	feet in the collar.
Total	9	8	

Turn in the dead-eyes, for a full due, by heaving the ends into their places (*inside* of the standing-part) with a strand of rope or jack-screw, and pass the quarter and end seizings. Now reeve the dead-eyes full, by the following method, which is the most economical, and always ensures a proper length: Place the coil, from which the laniards are to be cut, on deck amidships, on its edge, and take either end in each of the channels, beginning at the side-hole of the *lower* dead-eye, under the *standing*-part of the shroud, from *in, out,* and reeve the holes full; turn a Walker's* knot into the extremity of the laniard last rove, (which will be under its proper part of the shroud, viz. the *end,* so that in setting up the running part of the laniard brings the strain on th standing-part of the shroud,†) clap on the bight, render all parts, haul taut, and cut off from the coil as desirable. By adopting this method, a little time is saved, as two laniards can be cut and rove at the same instant from the same coil.

Now proceed to set up the lower-rigging, which should be done as follows:

Place canvas on the shroud about halfway up, to avoid chafe, and tail the single block of a luff over it. Turn a cat's-paw into the bight of the laniard, and to a toggle thrust through, hook the lower-block, (which is double,) and lead the fall up to the pendant-tackle, as in No. 12. Set up all the shrouds in this manner, a pair on each side at a time for a full due, (sending down the luffs,) and secure the end of the laniard by thrusting it through the space between the dead-eye and throat-seizing, around one part of the shroud, *inside* of both, and around the other part

*The *Matthew* is omitted, as being superfluous, for when it is considered that it is the only knot in use bearing a man's name, the *surname* is sufficient for identification.

†It appears that this theory (upon the correct observance of which so much stress is laid) has in part exploded, although but recently. On board of the ship "Minesota," the opposite of the rule was observed in the fitting of the port lower-rigging, (in order to make it correspond, in appearance, with that on the starboard-side;) and on questioning Captain A——— about the unseamanlike manner in which his laniards were rove, he replied, that "although the strain comes on the *running end* in *setting up*, yet after securing it reverts to the *standing-part,* and thus fulfills the conditions of the old rule; *without* causing the dead eye to slue, (as asserted by the advocates of the first method,) provided, of course, *it be turned properly.*" The proof of this was evident, even after the vessel had performed a voyage to Liverpool and back; for on trying the laniards, the standing-parts were taut, while the running ones were comparatively slack, and the dead-eyes perfectly square; Glascock, Gower, Griffith, and the host of *literary* seamen, to the contrary notwithstanding.

like the first; thence back, as it came out, racking the end to the *inner* and *outer* parts. This is done to prevent the dead-eye from turning out, in case the throat-seizing should snap, and also as a support to it.

No. 12

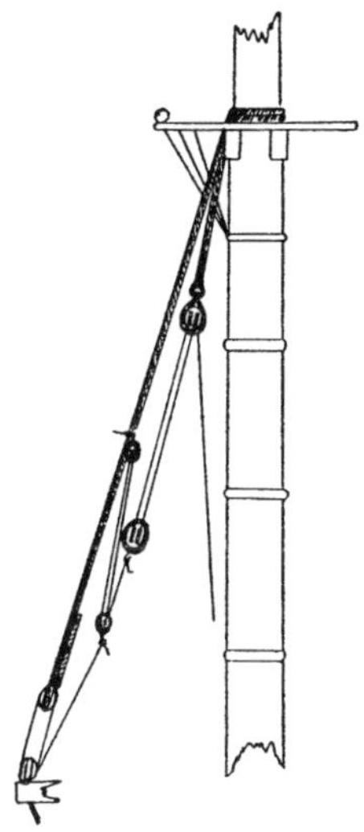

NOTE.—Dead-eyes are turned in as follows: Lay a strip of tarred parcelling in the score, and heave the shroud taut around the dead-eye (*with* the lay of the rope) by means of strands, (rove through the laniard-holes,) and a Spanish windlass. Reeve the first turn of the throat seizing through the eye in the end, around both parts of the shroud, and pass the other turns, usually *nine*, from the *end* towards the standing-part, hauling them merely *hand*-taut, so that when the shroud end is turned up, all may bear equally in the strain; thence back, underneath the turns, up through the eye, and pass the riders in the cutlines of the first, securing the end (which should be whipped) around the standing-part of the shroud.

The method of "cutter-stay fashion," although still holding many *theoretical* advocates, has but few *practical* experimenters; for scarcely a ship is seen now-a-days where this method of turning in the rigging is resorted to, on account of the chafe and premature decay induced by the shortness of the nip. But Glascock says, "The more this method is tried, the more serviceable it will be found;" and cites, in proof of the assertion, "that in the heavy hurricane which the squadron of Sir Richard Strachan encountered in the late war, the *only* ships that did *not* snap the throat-seizings of the lower-rigging, draw dead-eyes, or carry away spars, were those that had their rigging in cutter-stay fashion."

On board of the sloop-of-war "Fairfield," fitted in the summer of 1841, the method of cutter-stay was adopted, and found to answer all the conditions perfectly. But on her arrival in the Mediterranean the rigging required turning in afresh; and on examination it was discovered that owing to the neglect of the riggers in not putting a strip of tarred canvas around the *nip*, a very considerable chafe had been made, which rendered the turning up of the ends necessary, although, in justice to the system of cutter-stay, it should be added, that the present method was substituted on that account only.

Having secured the masts in the desired position, proceed to set up the stays by the following method, observing, first, to let the collars down snugly over the eyes of the lower-rigging, and to pass the remaining turns of the eye-lashing,—*not*, however, so as to bring the legs "eye to eye," as the stay can be set up afterwards *aloft*, (if the amount required be but slight,) without the trouble of casting off the seizings, and working on the reeving end.

No. 13

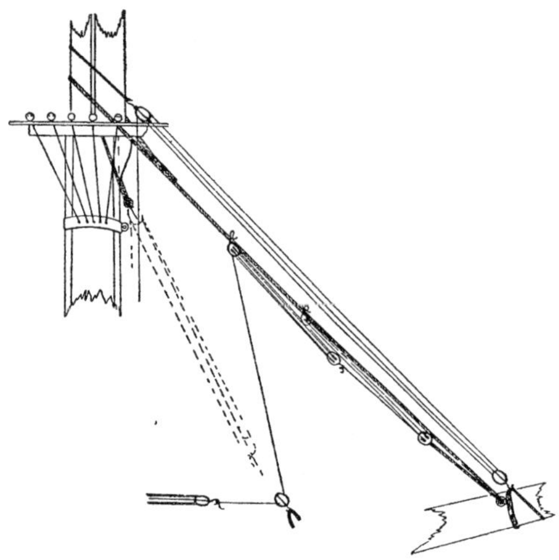

At the distance of eight or ten feet up the stay, clap on the single-block of a luff-tackle, (having canvas underneath to protect it from chafe,) and hook the lower one to a strap around the end of the stay, rove from *forward, aft*, through the thimble, (this supposes straps to be used.) Then to a backwall* in the hauling part, which leads upward, hook the *single* block of another luff, (the double one of which is tailed on to the stay higher up,) and lead the fall of this last-mentioned luff, down through a single-block on the bowsprit, in a line with the stay, and hook in a deck-purchase, or use the pendant-tackle, as illustrated in No. 13. Having got the stays taut, (which will be the case when they form a *straight* line,) pass the seizings, (like those used in forming the eye of a shroud,) and take off the luffs. (Plate 5, Fig. 22).

Now seize on the sheer-poles with a cross-seizing, to keep the dead-eyes from slueing, (on account of the tendency to unlay in the shrouds,) having a strip of tarred canvas, or (which looks neater) leather underneath, to prevent chafe, and prepare to rattle down. Or, should the old custom of futtock-staffs on the legs of the lower-rigging be still adhered to, (which, by-the-by, is an abominable antiquity,) swifter in the shroud-legs, by girting them, towards each other, with a piece of small stuff, but not *too* much, lest the seizings of the ratlins should part when let go; and seize the staffs, on *outside*, to all except the swifters, with their shrouds

*Brady calls it a "*blackwall.*"

and the cat-harpen legs, (the latter secured with cross-lashings, in the wake of each shroud and the staff,) the distance of half the width of the after-part of the top from the eye-seizing down. This system has, however, fallen into disuse; a far better and neater plan is to have iron rods, or rope legs, from a spindle-band around the mast, below the hounds, to the plates in the top rim, setting up, in the former case, with turn-buckles or screws; and in the latter, with a lashing, by means of a Spanish windlass or tackle, to the deck. This method is infinitely preferable to the one for which it is substituted; since, if the lower rigging becomes slack, the dead-eyes of the topmast rigging still remain fixed and stationary, which, in point of security, is a consideration of some magnitude.

Hook the futtock-shrouds to the plates in the top, observing to have the points of the hooks *in-board,* so that the bights of ropes from aloft, as studding-sail tacks, &c., are not in danger of being caught over them; and then proceed to rattle down, first lashing oars or spars athwart the rigging, at the distance of four or five feet apart, for the men to stand on while at work.* The ratlins are usually of eighteen-thread stuff, fitted with a small, neat, spliced eye, (thrust once and a half,) which is seized on to the shroud with marline, (or, which is neater still, a rope-yarn, twisted up the opposite way, and rubbed smooth,) at the distance of fifteen or sixteen inches below the preceding one. A clove-hitch is then formed *outside,* around the next leg, with a marlinspike, and the ratlin hove taut. In this manner it is made to reach the after-shroud, and then seized on as at the commencement; every fifth or "sheer" ratlin being extended to the swifters, which, with these exceptions, are omitted. (Plate 5, Figs. 20, 21, and 23).

An exceedingly simple *ruse,* (although perhaps unworthy a significant reference here,) where the ratlin is too long to be seized on thus, and yet not sufficiently so to allow a new eye to be formed, is to heave turns in with the lay of the stuff, until shortened up as much as required. This is vulgarly defined, by the *non leges scriptae* of nautical phraseology, an "Irish splice;" and notwithstanding its simplicity, is quite worthy of a stowage among the "wrinkles" of one's professional knowledge.

Now unreeve the girts employed in swiftering in, and black the rigging.† After which, send down the spars, and square and cap the ends.

NOTE.—The "full due" referred to in the preceding paragraphs is by no means intended as the only, or final setting up the rigging is to receive in the vessel's outfit.

*Prior to ratling down the rigging, a line parallel to that of the vessel's flotation should be drawn across the shroud-legs. (on both sides,) through the points where it is intended to seize on the lower ratlins, in order to have them correspond with her "sit" upon the water. And if these marks be continued upward to the tressle-trees, at a distance apart varying from fourteen to sixteen inches, as the rigger may choose, (that being the extreme separation,) the work can be greatly facilitated by commencing with three or four ends of the ratling stuff at the same time, in as many different places on the rigging, and without the necessity of refering constantly to the "measuring stick."

Nothing is here said of *neatness* in this particular, as it is supposed to be a consideration sufficiently regarded by every rigger; since no neglect is so palpable, even to the eye of the most inexperienced, as that of ratling down the rigging in a careless and slovenly manner.

†Various compositions have been used for this purpose. *Glascock* recommends "Stockholm tar, with a certain portion of salt-water, boiled together—and put on *hot.*" *Fordyce* says, "Stockholm tar, coal-tar, and salt-water in equal proportions, heated up in a fish-kettle." *Brady* suggests the following. "To half a barrel of tar, add six gallons of whiskey, four pounds of litharge, four pounds of lampblack, two buckets of boiling beef-pickle, or hot salt-water; mix well together and apply immediately."

The season of the year—the length of time in port, and the celerity, or tardiness of equipment, are all circumstances which either make it necessary to give the rigging other *pulls*, or to afford opportunities for doing so. *Fordyce*, thus states the criterion of the British service:

"Forty-eight hours, or so, after placing, setting up, and swiftering in the rigging, if the weather be favorable, get another pull of it, and mark off for squaring.

"Five or six days after this, get a third pull, turning in the dead-eyes, with their proper seizings to the squaring marks, and bringing them down for good and all. Rattle down the rigging.

"Square off and point, or cap, (with canvas or leather,) the ends of the shrouds, and point, or whip the ends of the laniards.

"Another, or fourth pull of the rigging, (supposing it to be new,) a fortnight or so after the third, will amply prepare it for the sea, if the weather prove favorable."

This system may answer very well for Navy Yards, where time is not considered, but it is totally incompatible with the enterprise and dispatch characteristic of the American mercantile marine. In that branch of the service, two settings up are generally given at first, and when every thing is *a taunto*, just before going to sea, it is usual to get a pull of the laniards and to square the ratlins, etc.

The routine of rigging embraced in the foregoing article refers exclusively to the foremast: yet the difference in the other masts is so immaterial that it is only necessary to allude to a few peculiarities. Over the mainmast-head, no clump-blocks and straps are placed, and the mizen lower-pendant, which is single, is fitted with a cut-splice, while the odd shrouds (although disapproved of by seamen generally) are fitted with horse-shoe eyes (Plate 5, Fig. 24). In vessels of war the starboard main-stay should be passed *over and under* the bitts in setting up, and the port one *vice versa*. This *ruse* will greatly diminish the chances of losing both by a single shot in the event of action. (Plate 6, Fig. 26).

TOPMASTS.

The operation of getting in topmasts, in lieu of being confined to one particular method, is performed differently by almost every rigger. But the established limits of the work preclude the introduction of more than one, and render it necessary on our part, to make what might be *critically* termed an "invidious distinction." The following method, adopted by an officer of the navy, whose abilities as a seaman are well known to the service, is selected, on account of its perfect feasibility under all circumstances, as the best and simplest. Acting upon the supposition, assumed in the routine of rigging explained in the foregoing pages, that the ship is moored in the stream, tow the mast alongside with the *head* forward,* having pre-

*Captain ——— recommends in all cases of getting spars on board while in the stream, (where there is a *tide*-way,) to tow them alongside in the reverse manner—a topmast, for example, with the *heel* forward, &c.; then by letting the foremast end be swept aft by the current during the operation of swaying it on board, there will be no necessity for a hauling-guy to rouse the heel forward *against* the tide.

viously secured a large bull's-eye, or single block, to the hounds on each side, in the same plane with the lower sheeve-hole.

No. 14

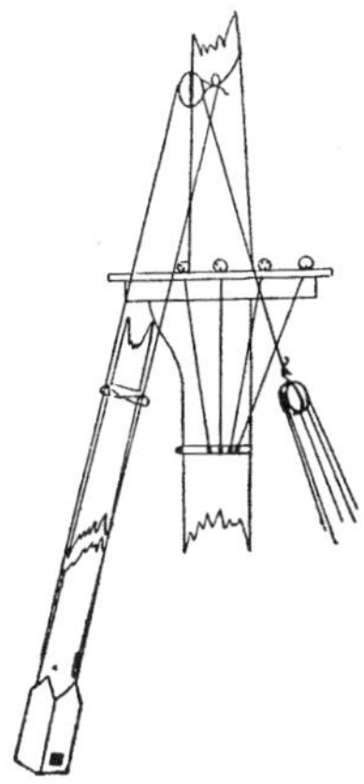

Hitch the end of a hawser at the mast-head, above the eyes of the rigging, (*outside* and clear of the top,) and reeve the other through one of the bull's-eyes, (or blocks,) and the sheeve-hole; thence up through the opposite bull's-eye, and a block lashed at the mast-head, (again *outside* of the top,) as in No. 14; leading it to the deck, and clapping on a pendant-tackle. With this purchase, and by means of easing-guys and jiggers, sway the mast in-board, and up and down the lower-mast; lashing the head to the crotch of the fore-stay collar,* and resting the heel on deck, in order to shift the mast-head-block, and standing-part of hawser lower down, to lead through the lubber's-hole. Remove the deck scuttle—cast off the stay-lashing, and lower the mast, until clear of the top rim; then sway it up, and point it through the tressle-trees and round-hole of lower cap; which latter is sent up "before all," with the girtlines immediately after rigging the lower-mast, by bending them on through the round-hole, and stopping them along to the after-part, observing to keep the bolts *uppermost*, so that they do not come in contact with the top-rim, &c., in the cap's passage aloft. Reeve a stout strap through the fid-hole, to which, hook both the pendant-tackles; take off the bull's-eyes (or blocks) at the hounds, and mast-head—unreeve the hawser, and prepare for shipping the CAP, which is done as follows: Pass secure lashings through the eye-bolts, over the topmast-head, (or *wedge* the cap,) and sway up on the pendant-tackles, until clear of the tenon of the lower-mast; then slue the heel of the topmast, by means of guys bent to the extremities of a long heaver placed in the fid-hole, until the cap is fair for shipping. White-lead the tenon and mortice; send up the cap-shore,† (to which

*Supposing it to be the fore-topmast.

†Cap-shores in the merchant service are now seldom if ever met with; the necessity of them being obviated in a great measure by the shortness of the cap, forward. But this *economical* dispensation is frequently attended with an experience of time and labor, in striking topmasts, on account of the weight of the spars and rigging bringing a heavy strain on the fore-part of the cap, "which," as Glascock justly observes, "causes it to droop and bind the mast."

a laniard should be attached for the purpose of securing it aloft,) and lower away, beating the cap down in its place, and tacking a piece of sheet-lead over as a protection from the weather.

Cast off the lashings, (or remove the wedges,) and sway the topmast-head a few feet above the cap. Lash a couple of stout burton-blocks to the tenon (having the falls rove,) which should be sent down "behind all" for the cross-trees, (placed on deck *well* abaft the mast.) Secure the lower blocks to each after-horn of the tressle-trees on the *upper* side, and stop the standing-parts along to the forward-horns, in the same manner as that resorted to in sending up the lower tressle-trees; having a guy from the mainmast-head, (if the fore-topmast cross-trees,) to keep them clear of the top in going aloft. Sway up on the burtons, bear off, cut the stops as necessary, and land them on the lower cap, where they should be securely lashed—having the forward part inclined upward, with the chock resting against the topmast. Cast off the burtons—remove the blocks from the tenon—or, if girtlines are used to get the cross-trees aloft, (as is sometimes done,) shift them at once to the middle-horns, ready for the rigging—white-lead all parts in the wake of each other, and lower away on the pendant-tackles, until the cross-trees come fair over the mast-head, cutting them forward, or aft, as may be necessary.

Of late years a new parrel has been introduced in the merchant-service, which causes a slight deviation in the routine of rigging, on account of its going *over* the topmast-head *prior* to the cross-trees, in order to avoid the difficulties, which would be otherwise experienced by the coopers and blacksmiths in completing it, if left until afterwards.

No. 15.

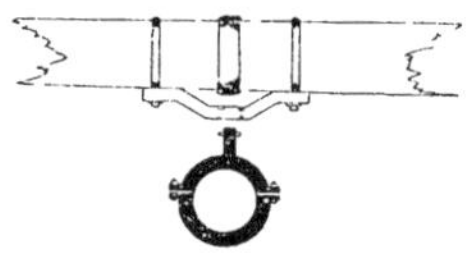

This parrel consists of a number of stout staves, forming a circle, (somewhat larger in diameter than the topmast,) confined by a broad iron spider-band, having a neck, or shoulder, which connects with a truss on the topsail-yard, by means of a key, as in No. 15. They were used on board of the store-ship "Southampton," in 1846; and although she encountered a series of heavy gales, on her return from the coast of Africa, after an absence of nearly two years, no accident happened to the yards whatever. The advantages which this parrel possesses over all the others, viz.: of keeping the yard clear of the rigging, and the facility of clewing down quickly, have been deemed sufficient to warrant an almost universal adoption of it, in the mercantile marine. It is perhaps well, however, to remark, that in order to check the play of the parrel, (which would otherwise follow,) the diameter of the topmast should not vary from the cap to the hounds. (Plate 6, Fig. 27).

Now sway up, and lodge the cross-trees on the hounds of the topmast, prying up the after end, and beating them down in their places. Hook the top-blocks,*

*The manner of doing which has always been a favorite "catch question" with the saline sages of the forecastle in their teachings of the "green 'uns;" on account of the emphatic "from forward, aft, *of course*," invariably returned to the query, or the equally prompt conjectural vice-versa, "*to be sure*' Neither of which is right—the block being hooked from *in, out, necessarily.*

and reeve the pendants, by passing each pointed end through its respective block, and sheeve in the heel, and clinching it *over the cap* on the opposite side, and *not* "to the eye-bolts;" after which, hook the top-tackles to the other ends, (fitted with thimbles,) and remove the fid-strap and pendant-tackles used in pointing the top-mast. Attach a girtline-block (vide page 30) to each (*middle* if three, otherwise) after-horn of the cross-trees, and prepare for rigging. Send up and lash the bolsters, well protected with tarred parcelling, and bolt the *gin-bar* down securely in its place. This is a late invention, and in addition to its perfect security, tends much to the neat appearance of the mast-head, by dispensing with the huge strap and pendant for gins and burtons, formerly in use. It consists of a short, stout bar of iron, placed across the topmast tressle-trees, between the "doublings" of the top-gallant-mast, with links for the gin-blocks (Plate 7, Fig. 28). The burton-pendant (now unknown in the merchant service) being fitted with a hook in the upper end, which connects to an eye-bolt on the *under* side of the tressle-trees. This plan was adopted on board of the "Columbus," 74, while flag-ship of the East India squadron.

No. 16

Objections are frequently made by seamen to rigging the topmasts before fidding, on account of the increased weight thereby brought upon the top-pendants, which (considering the advantages of safety, and a top unincumbered with rope) are not without some shadow of reason; and since it is usual to fid the mast first, in order to ascertain its fit, etc. before rigging, the method should be adopted. Having then fidded the topmast, proceed to get the shrouds and backsays up and over, in the following order, after the manner employed in LOWER-RIGGING, observing to have an eye-girtline at the tenon:

No. 17

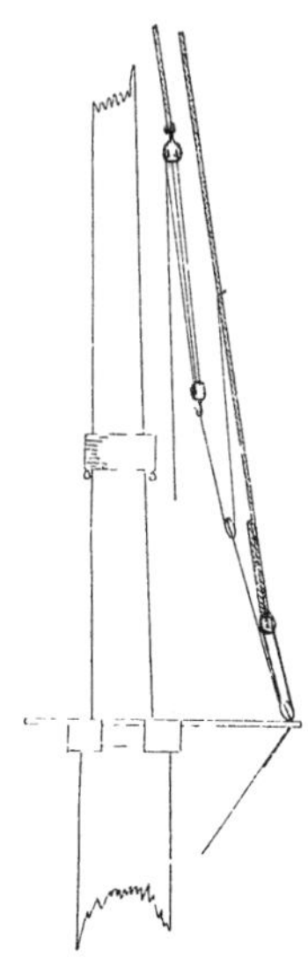

First. Starboard and port shrouds alternately—the odd ones being fitted with horse-shoes eyes.

Second. Standing backstays.

Third. Fore-and-aft stays.

Lastly. Breast-backstays, fitted with a neat eye, through which a *hide* grommet is formed above the eyes of the rigging, or a strap and toggle passed as recommended on page 12. Finally, send up the jib-stay, (if the boom be rigged out at the same time,) and reeve the topmast-stays through the bees on the bowsprit, clapping luffs on them, to give the mast the proper stay. Then proceed to set up the rigging by means of a runner and burton fitted as follows.

The runner (which has a thimble in one end) being rove through a single block, hooked to a cat's-paw in the laniard, is tailed on to the shroud-leg about six or eight feet above, and the lower block of the top-burton then hooked to the thimble, as in the annexed figure. Having set up the shrouds by means of this purchase in a manner precisely similar to that adopted in the case of the lower-rigging, turn the dead-eyes in to the standing backstays and set them up also. Lash on the sheer-poles and futtock-staffs* (Plate 7, Fig. 31)—the latter *inside*

*Which should be well parcelled and served, in order not to chafe the top-gallant rigging, as they are only used for it to lead over, in its course to the top.

of the topmast shrouds, and swifter in. Seize the forward cat-harpen legs* (on each side) to the forward shrouds, and the after ones (around the mast) to the after shrouds—then spar, and rattle down, and cap, and square the rigging. As a *finale,* reeve and fit the netting on the top-railing, and sway aloft the upper cap, which should be placed in the fore part of the topmast cross-trees ready for shipping (Plate 7, Fig. 29). Lastly, send down the girtlines from the tenon and horns, used in getting the rigging over.

NOTE.—Both futtock-staffs, and cat-harpen legs have disappeared from the merchant service (Plate 7, Fig. 31). The top-gallant shrouds setting up through bull's-eyes at the horns of the cross-trees, (Plate 7, Fig. 30,) which are retained by iron legs, with screws or turnbuckles, to a band just below the hounds of the topmast, like the futtocks of the lower-rigging.

N. B.—In many merchant vessels, and particularly those of small tonnage, the topmast rigging is simply rove through large bull's-eyes, instead of being turned in with dead-eyes and laniards. And frequently, after the lower and topmasts are rigged, a tarpaulin coat fitting snugly, is placed over the eyes of the rigging, as a protection from dampness and wet. This answers very well, and if painted white, does not detract from the neat appearance of the mast-head.

JIB-BOOM.

Almost every rigger has a particular method of getting this spar on board. *Brady* recommends using "the hawser, or tackle, which was left at the foremast-head, when getting on board the fore-topmast." *Totten* suggests hoisting it up "by one of the pendant-tackles, with the outer end forward;" while others employ "a purchase on the fore-topmast backstay." But the following method, adopted by the late Boatswain O'Neal, (and somewhat more explicit than either of the above,) in refitting the frigate "Cumberland," is thought to be the best. (Plate 11, Fig. 65).

Bring the boom alongside with the head forward, and reeve a spare piece of rope, (studding-sail halliards if at hand,) through the sheeve-holes in each end, a sufficient number of times, and make it fast. Overhaul down the main pendant-tackle, and hook it into a cuckold's-neck (Plate 11, Fig. 62,) formed in the bight of the span, having the boom to hank slightly *heel* heavy. Sway it up, bearing it clear of the ship's side—ease it in-board, and land it in the gangway; unreeve the span, and carry the boom forward, pointing it through the bowsprit-cap, and reeve the heel-rope, which is done as follows: Pass one end through a single-block, hooked to an eye-bolt on one side of the bowsprit-cap; thence through the sheeve in the heel, and clinch it to the other bolt, on the opposite side of the cap. Man the heel-rope, and rig the boom out, until the shoulders are just forward of he bowsprit end; then tar them in wake of the rigging, and get over the *foot-ropes first,*† which are fitted with neat-spliced eyes, (Plate 11, Fig. 63,) in the outer ends, (thrust once

*"They should be fitted short enough to admit of being frequently set up, before they becomes so much stretched as to make it necessary to send them down to alter"

†This is done, in order that the men employed in fitting or stripping ship, may have something to stand on, while placing, or removing the other rigging.

*This plan is preferable to the "cut-splice," which is rendered round the boom by the weight of the men, first on one side and then the other—a circumstance inducing chafe, and causing the splices to be forever *"lop-sided."*

and a half,) and wormed and served.* The strands are then unlayed, up to the service, and Walker's knots worked in at equal distances, (to keep the men from slipping, on account of the steeve of the boom;) after which, the rope is again laid up, and the inner ends formed into artificial eyes, which are seized to bolts in the bowsprit-cap. Thus fitted, they should be long enough to allow the men who go out on the boom, to stand with the lower part of their breasts against it. Next put over the *jib-guys* fitted with a "cut-eye," (Plate 11, Fig. 64,) formed by seizings on each side, and covered with canvas. Lastly, put over the eye of the *martingale-stay*, made in the same manner as those for the foot-ropes. Reeve the jib-stay through the hanks and the inner sheeve-hole of the boom-end; and sway the dolphin-striker up by means of a tackle from the bowsprit-cap, hooking it to its respective eye-bolt, and putting over the lower eye of martingale-stay, and the gob-lines, fitted with a round-eye in one end, and a hook-thimble in the other, which latter, connects with a bolt on each side of the cutwater. Finally, ship the wythe for the flying-jib-boom (Plate 11, Fig. 66); man the heel-rope, and rig out, placing the heel in the saddle, and clamping it. Pass the belly-lashing, set up the jib, and martingale-stays, seize the top-gallant bowline-blocks to the jib-guys, (far enough out from the shoulders to allow the flying jib-boom to pass clear,) and unreeve the heel-rope.

NOTE.—On board of merchant vessels an iron band is placed around the shoulders of the jib-boom, having an eye-bolt on each side for the guys and foot-ropes, (fitted with hook-thimbles, (Plate 11, Fig. 61,) and one on the under side for the martingale, which connects with a hook also. The gob-lines and martingale, which are made of chain, connect to the dolphin-striker in a similar manner (Plate 11, Fig. 60).

TOP-GALLANT-MASTS.

Lash the jack-block aloft above the eyes of the topmast rigging, and reeve the top-gallant-mast rope: first through the block, then the square hole in the cross-trees, and lastly the lizard (see No. 18) at the royal-mast-head, and the sheeve in the heel, seizing it to the hounds of the top-gallant mast, and leaving end enough over to clinch to the eye-bolt in the upper cap, (when shipped,) before fidding. Send the hauling-part down through the cross-trees and lubber's-hole to the deck, and rack both parts of the mast-rope securely together just above the lower sheeve-hole, and sway aloft, bending a tripping-line on through the fid-hole to guy the mast clear in its passage up. Point the head of the royal-mast through the cross-trees—take off the lizard, (which is now of no further use,) and prepare for shipping the upper cap. Sway the mast up through the latter—placed *above* the collars of the stays, and secure it to the hounds of the royal-mast by means of wedges and lashings. Send aloft the cap-shore—sway higher, and proceed after the manner directed in shipping the lower cap, (page 29).

Remove the lashing and wedges, and clinch the end of the mast-rope to the eye-bolt in the cap, slacking down the former, (kept from rendering by the rackings,) in order to take the strain from the jack-block, which must now be unlashed and hooked to its proper eye-bolt also: after which, cut the racking and the seizing at the hounds, letting the mast hang by the mast-rope, the end of which is belayed on deck Lash a stout girtline-block to the cap, and overhaul the line down abaft all,

No. 18

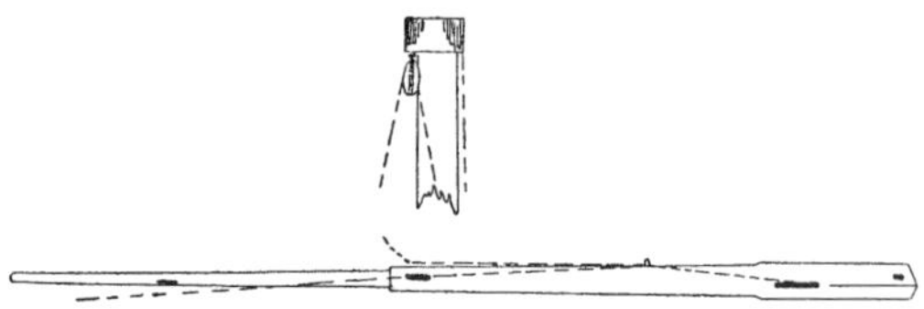

for the jack and funnel,* which should be placed on deck, with the rigging over in the following order, the eyes being made to fit the cylinder exactly, and covered with canvas.†

First. Fore-and-aft stay.
Second. Shrouds.
Third. Backstays.
Lastly. Breast-backstays, (over the *fore*, last of all, flying-jib stay.)

Bend the girtline to the rigging around all parts, the length of the *topmast*-head below the jack, and put a good stop through the cylinder, sending it aloft, with a guy attached, to keep it clear. Cut the stop when high enough, and place the jack over the cap, assisting to do so, if necessary, with a rope's end through the royal sheeve-hole. Now lower the top-gallant-mast, cutting the cylinder forward, until it is fair over the hole, and lash it, for fear of accident. Reeve the top-gallant-shrouds through the horns of the cross-trees, (or bull's-eyes, if in the merchant-service,) and place the royal-rigging (sent up by the girtline also) over the funnel. Sway up on the mast-rope—point the royal-mast-head through—ship the truck, (with signal halliards rove, spindle and conductor‡ attached,) and fit the *royal-stay*

*The rim of which should be rounded off, or filled in with lead, and a circular piece of canvas placed over to prevent chafe.

†The exclusion of odd shrouds and breast-backstays from the top-gallant-masts in the merchant service renders the funnel here referred to *dispensable*—a single grommet being used instead. This, however, is generally fitted with a small becket or thimble, on each side, (a little forward of the centre,) for the top-gallant-lifts to seize to, which are made to go *standing*, of just sufficient length to allow the yard to hang clear of the upper cap when clewed down. (Vide TOP-GALLANT-YARDS.)

‡It is a matter of no little surprise that so many vessels, in the merchant service, are equipped without this most *necessary* appendage; and that even those used in the navy are little else than *nominal* conductors, put up apparently in view of keeping within the pale of some *expiring* regulation. A great number of the vessels, from which no accounts have ever been received, have doubtless been lost *solely*, in consequence of not being properly protected against the effects of *lightning*. That imperfect building, or deficient navigation, are the causes of all the calamities at sea, no one who is acquainted with the characteristics of the American mercantile marine, will attempt to maintain for an instant; or, that all of those names are written in the dark and mysterious calendar of "lost at sea," have *necessarily* gone down in "tempest-tossed" and foundering ships. The effects of lightning, as witnessed on shore, are often dreadful indeed, but when the horrors of a conflagration at sea are superadded, without the hope of escape, or rescue, no pen can portray the agony of such a scene; and it seems almost madness to neglect the *only* means of guarding a ship against an element so fickle, and yet withal so fatal in its effects and consequences.

down on the hounds, (above, which a grommet* is sometimes worked;) then the *shrouds*, and lastly, the *backstays*. Reeve the shrouds through the jacks, the backstays to the top and channels, and the fore-and-aft stays forward, through single blocks at the jib-boom end, (if the flying-jib-boom should not be rigged out, at the same time.) Cast off the funnel-lashing from the cap—man the mast-rope—unbend the tripping-line—sway up, and fid, setting up the rigging, *temporarily*, to steady and secure the mast.

FLYING-JIB-BOOM.

Sway it on board, either by means of a bill-tackle on the topmast-backstay, or use the span, as directed under the head of JIB-BOOM, and get it ready for going out. Lash a single block to the jib-stay, (and *not* "to the wythe," as sometimes directed, for if so, the boom cannot be rigged out at all,) through which reeve the heel-rope, taking a hitch with the end over the heel of the boom, and clapping stops around both, to the shoulders. Pass *slip*-ropes also over jib, and flying-jib-boom, (to hold the latter up, in its passage out,) and rouse on the heel-rope, pointing the end of the boom through the wythe, until the shoulders are clear of the jib-boom end. Tar them in the wake of the rigging, and put over a *grommet*, with thimbles on each side, for the royal-bowlines (Plate 8, Fig. 33). Next, the *foot-ropes*, (fitted like those for the jib-boom,) and seize the inner eyes to the jib-guys; then the *flying-jib guys*, and lastly, the *martingale-stay*, which reeves *through* the dolphin-striker, and leads to the head. Reeve the flying-jib, and fore-royal-stays, (passing the former through the hanks;†) cut the stops; rig out, and pass the heel and belly lashings. Set up the martingale, and flying-jib-stays—take off the heel-rope—set up the royal shrouds and backstays, and finally, the fore-and-aft stays, for a full-due. (Plate 8, Figs. 32, 34 and 35).

NOTE.—In the merchant service, the martingale-stay is frequently of chain, (Plate 8, Fig. 36,) reaching only to the dolphin-striker, where it is secured by

Some years since a lightning conductor was invented by Capt Harris of the British Navy, and subsequently adopted throughout that service. This consists of a continuous line of double plates of *copper* (38 oz. per square foot) let into a groove on the after part of the masts, and extending from the truck-spindle to the lower cap, gradually increasing in width from two and a half to four inches. At the cap, the conductor meets a similar line of metal, running entirely through the centre of the lower mast, which is stepped upon a broad plate of copper on the kelson—from whence the fluid is discharged by means of bolts of copper running through and through at each extremity of the plate. The advantage of this method consists in the fact of its being a better conductor of lightning than iron in the ratio of 6 to 1; and by being made of *plates*, in lieu of rods or links, it exposes *four* surfaces to the action of the fluid besides imparting to the lighter and loftier masts a very considerable degree of support. When we consider that not a single war-vessel in this country is fitted with one of these conductors, it amounts almost to a stigma upon the American navy.

In reference to the merchant service, it would seem that the story of the ill-fated "Logan," as detailed in the "History of Remarkable Shipwrecks," has failed to awaken attention to this subject; and since the *moral* of a hundred tales equally frightful cannot, or *will* not be appreciated, it would not only be a *judicious* adoption, on the part of *insurance companies*, but a *philanthropic* one, also, were they to compel eevry ship to be fitted with *some* means of guarding her against the chance of loss

*In the merchant service this grommet is fitted with brackets, or thimbles, for the royal lifts.

†Vide BENDING SAILS.

means of a hook to an iron band with eye-bolts; and the foot-ropes and guys fitted with hook-thimbles in the outer ends, which connect to the shoulder-band in a similar manner.

On board of the frigate "Cumberland," in lieu of a grommet and thimbles, for the royal-bowlines, small bull's-eyes were seized to the guys for that purpose; and in rigging ship, the royal shrouds and backstays were set up *first,* because the weight of the stays alone brought the mast forward too much.

TOPSAIL-YARDS.

All the *writers* on seamanship, place topsail-yards next in order after lower ones. in rigging ship; but this routine is not *always* observed in practice, and particularly where expedition is required in the equipment, on account of the increased labor and difficulty of crossing, if the lower yard is rigged and sent up first. This consideration, although apparently a slight one in the abstract, nevertheless becomes of some magnitude in the practical application of this *ci-devant* rule. Having towed the yard off to the ship, (supposing her to be in the stream,) say on the port side with the starboard yard-arm forward,* lash a large single-block at the topmast-head, having a strap sufficiently long to permit it to hang clear of the tressle-trees Through this reeve a hawser-down, (outside of all,) and bend it on the slings of the yard, either stopping it to the forward (in this case starboard) quarter, with stout seizings, or use a lizard, and secure the ship's side from chafe by fenders and skids. Hook the port pendant-tackle, also, to a strap on the after-quarter, and man it with the hawser, (taken to the capstan,) swaying the yard on board, which must be kept from canting aft against the mast, by means of a purchase, or guy, leading from forward. Ease the lizard (or stops) as necessary—sway on the pendant-tackle, until clear of the ship's side, and lower away, landing the yard as you had it alongside, (viz. with the starboard yard-arm forward.) in the port gangway, on chocks, which should also be placed underneath the *inner* quarters, to keep the yard from becoming bowed in the slings through its own weight. Now cast off the hawser and tackle; tar the yard-arms, and prepare for rigging (Plate 9, Fig. 37; Plate 10, Fig. 51).

1. Tye-blocks (Plate 9, Figs. 38, 39, and Plate 10. Fig. 52.—Although invariably iron-strapped, and connected by means of a bolt to a band going around the slings, (after the yard is crossed,) it is thought proper to describe the manner of securing the block, when fitted in the old-fashioned way; as the present method, though by far the best and neatest, is nevertheless not entirely free from accident. Having leathered the slings, find the exact centre of the yard,† and heave the block into the strap, (formed of a long and short grommet, wormed, parcelled, and served,) with a strand and Spanish windlass, and seize it *loosely;* after which, lash it on the yard with the breech fitting down snugly, and the sheave standing fore and aft, passing the rose-seizing, connecting the eye of the long

*Vide note, page 28, in reference to getting spars on board.

†Where two tyes are used, as on board of heavy ships, the blocks are separated from each other by a space equal to the diameter of the topmast.

grommet (going around the yard) to that of the short one, on the forward side, underneath. N. B.—On board of many merchant-men (of small tonnage) this block is dispensed with, and the tye hooked to an eye-bolt in the sling-band.

2. PARREL (Plate 9, Figs. 40, 41).—If the newly-invented parrel, described under the head of TOPMAST, should not be allowed, prepare to lash the rope one (fitted as follows) on to the yard. This consists of a long and short leg, marled together and leathered, of sufficient length to go around the mast and yard on each side. An eye, which is wormed, parcelled, and served, is spliced in the ends of the two legs, and stout quarter-seizings put on both, close to the eyes of the short leg. The long one is then passed around the quarter of the yard, just half the diameter of the topmast from the centre, and secured to the short one by a rose-seizing on the *upper* forward side. When the yard is crossed, the remaining leg is passed on the opposite side, and secured in the same manner.

In many instances in the merchant service, *jaws* (Plate 10, Fig. 59) with a simple lashing abaft the mast have been substituted for both the parrels here alluded to, and are preferred by many seamen on account of their cheapness and durability.

3. QUARTER-BLOCK: (Plate 9, Fig. 42 and Plate 10, Fig. 53) should always be iron strapped, on account of the facility with which the yards can be shifted, without the necessity of unreeving sheets and clew-lines. This plan has been universally adopted in the merchant service;* but it is a subject of regret, that a method tending so much to rapidity of movement, in the frequent and *necessary* evolution of shifting yards, should not have found its way into the navy long ere this. When the old plan of rope is resorted to, they are fitted with single straps, which are served—the *after* part only being leathered, to prevent chafe, and secured to the yard by a rose-seizing on top, having a span-lashing underneath, to keep them from spreading.

4. STRAPS AND THIMBLES, (Plate 9, Fig. 43,) (fitted with small eyes,) for standing-parts of topsail clew-lines. These are simply served and secured to the yard on top, (close to the quarter-blocks,) with rose-seizings.

5. ROLLING-TACKLE AND BURTON-STRAPS.† (Plate 9, Fig. 44.)—These are formed of grommets, doubled, (*not double-grommets*,) and thimbles‡—served and secured with rose-seizings on the forward side—the former, about one-third the length of the yard-arm out from the slings, and the latter just inside of the sheeve-holes.

6. HEAD EARING STRAPS AND THIMBLES (Plate 9, Fig. 45.)—These are fitted in a variety of ways—sometimes of *sennet*, and nailed around the yard, as done on board of the frigate "United States" in 1846; or *in one* with the eyes of the jack-

*The quarter-blocks on board of merchant vessels are *threefold*, in order to receive the reef-tackle, which comes in along the yard underneath, in lieu of reeving through the outer sheeve to the burton, as done in the navy. This is a good plan, as it serves the purpose of a down-haul-tackle in clewing down.

†These are frequently omitted in the merchant service. The former, particularly on board of those vessels fitted with the new-fashioned parrel, which, in a measure, does away with the necessity of rolling-tackles, on account of the little play given to the yard. The latter is dispensed with, for "*causes unknown*."

‡The reason is obvious—for, in the event of parting the strap, it could be replaced without disturbing any of the other rigging, which would be impossible with a simple grommet going *over* the yard.

stays and foot-ropes, as in the "Cumberland" during her late cruise in the Mediterranean. And again, in the merchant service, a simple eye-bolt in the shoulder-band of the yard is made to answer the purpose. The first method is preferable on account of forming a "soft bed" for the foot-ropes. The second for neatness—although there is a sacrifice of security, which is hardly commendable; while the third has advantages over both in point of strength.

NOTE.—This routine of rigging may be objected to by the sailors of "auld lang syne," as reversing somewhat the "ancient order;" but it would be well to remember that unless the straps for the head-earing be put over the yard-arm *first*, it will not only chafe the jack-stay, but *necessarily* every thing else *inside* of it.

7. JACK-STAYS (Plate 9, Fig. 50).—Wood and iron are used throughout the merchant service; but where rope is preferred, they are served the whole length, and fitted to go over the yard-arms with eyes; after which, they are rove through staples on the upper forward side, (being protected from chafe by strips of leather,*) and set up to each other in the slings, by means of small eye-lashings.

8. FOOT-ROPES.—These are fitted with an eye, going over the yard-arm abaft, which is merely wormed, and the splice served. They are then rove through the stirrup,† and the ends (which are served as far out as they are liable to chafe from the topmast rigging) taken abaft the mast, (*when the yard is crossed,*) and secured to the opposite quarters on *top*, by means of an eye-lashing, passed over the yard—round on the forward side—underneath, up, and back through the eye again, a sufficient number of times; after which, two half-hitches are taken around all parts to secure the end. This plan of leading them is recommended, on account of the facility with which the men can get on and off the yard. In merchant vessels, and on board of the store-ships, the foot-ropes are fitted with a hook, which connects to an eye-bolt on the after-side of the shoulder-band (Plate 10, Fig. 58; Plate 14, Fig. 92).

9. BRACE-BLOCKS (Plate 9, Fig. 47 and Plate 10. Fig. 54).—These are fitted variously. In the merchant service they are attached to the shoulder-band on the yard, by means of a hook-thimble seized into the strap of the block. But in the navy two straps are used, connected with lock-thimbles, and fitted as follows. The *yard-strap*, into which one thimble is seized, consists of a grommet, wormed and covered with canvas, and fitting so snugly as to require to be driven on. The *block-strap* (a double one) is cut the necessary length—the ends spliced together, and then wormed and covered with canvas. After which, the bight is thrust through the other thimble, and both parts passed around the block, which is secured into the strap by a round seizing, between the breech and thimble

10. LIFTS.—In large vessels, as ships- of-the-line, they are fitted double, reeving through a block (Plate 9, Fig. 48 and Plate 10, Fig. 55) over the yard-arm next after the brace; but in the smaller vessels of the naval service, they are formed

*In rigging, as little leather as possible should be used, on account of its tendency to produce rot; just sufficient to guard against chafe, being deemed by men of experience a judicious use

†Which are formed with thimbles in the lower ends, and small artificial eyes in the upper ones, seized to the jack-stay staples.

with a simple eye, which, with the splice, is wormed and covered with canvas; while those for merchant-ships are fitted with a hook-thimble to the shoulder-band.

11. FLEMISH-HORSES (Plate 10, Fig. 57).—These are spliced around the thimble on the pacific-iron for that purpose, and the eye in the other end secured on top of the yard to the jack-stay, the length of the shoulder inside of the sheeve-hole, with a rose-seizing.

12. JEWEL-BLOCKS (Plate 9, Fig. 49 and Plate 10, Fig. 56), which are single and rope-strapped, are sometimes secured to a thimble around the pacific-iron, outside of that for the flemish-horse, and frequently fitted to go *over* the yard-arm; but the best plan (as it admits of their being taken off at all times without difficulty) is to connect them to the pacific-iron with sister-hooks. (Plate 9, Fig. 46).

13. TYES.—Where rope is used, (which, by-the-by, is not the case in the merchant service, chain being universally adopted,) they should be fitted of sufficient length to admit of sending the yard up and down with them—stopping the surplus end of the standing-part, (which is secured at the mast-head, "breeching-fashion," with two stout seizings,) down the forward-swifter of the topmast rigging. This plan is also recommended by Glascock and others, for vessels of war; as it saves both time and trouble in "shifting," and the end serves to sling the yard with, in the event of action.

Lastly. See that the armorers and carpenters have prepared the yard, by attaching the boom-irons, reefing-cleats, bolsters (inside of sheeve-hole) for top-gallant-sheets, etc., and prepare for crossing.

As on all nice points of seamanship, there are many opinions in reference to sending up topsail-yards; but the following method, attested by practice, and combining as it were the commendable features of almost every other system is selected as the best: Reeve the hawser, used in getting the yard on board, through the thimble of a stout lizard, (secured in the starboard rolling-tackle strap,) and a large single block lashed on top of the yard, the diameter of the topmast outside of the tye-block, (the starboard one, if two are used,) clinching the end aloft, above the eyes of the rigging, and sending the other through the lubber's-hole to the deck. Coil the lifts on the quarters of the yard, (stopping them to the jack-stays,) and reeve the braces with marrying-lines, observing to dip the starboard (or *upper*) one *over* the lower stay. Overhaul the top-burtons from aloft, and hook them to the yard-arms; as also a fore-and-aft tackle to the slings to keep the yard from chafing against the mast, as it goes up. Now man the hawser and starboard burton, and sway away, carrying the *port* (or lower) arm aft, and placing a mat under, to prevent injury to the deck. As soon as the yard is clear of the lower-stay, ease the lizard, and commence crossing—keeping to the slack of the fore-and-aft tackle, and hauling on the lower burton and starboard brace. Reeve the lifts through the sister-blocks, and as the yards rise above the lower cap, square it; bring to, and pass the parrel, or *truss* the yard, as the case may be, sending down the fore-and-aft tackle. Now reeve the tyes (13) through the gins, (or mast-head sheeve;) splice in a single block to the reeving end, reeve the runner, and hook the fly-block with the halliards rove,* taking the strain from the burtons and hawser, which may be sent

*Vide RUNNING-RIGGING.

down, together with the yard-block and lizard; observing, as a *finale*, to place a small square piece of wood between the slings and lower cap, to keep the yard from bowing, in case the halliards should be slacked, or let go.

N. B.—This routine supposes the yard to be lying in the *port* gangway, with the *starboard* yard-arm *forward.*

LOWER YARDS.

In order to illustrate under one explanation all the principles involved in the operation of getting the yards on board, it will be necessary to grant the premise, assumed with regard to the position of the ship, viz.: at anchor in a *tideway.* Having lashed a stout block, with a hawser rove to the crotch of the lower-stay collar, (supported by a burton from the topmast-head,) tow the yard off to the ship, say on the starboard side, with the *port* yard-arm *aft,** and secure a large single block to the slings, as also a bull's-eye on each side of the *port* quarter, well out on the yard. Through these reeve one end of the hawser, and clinch it over the lower cap, sending the other part through a leader on deck to the capstan (Plate 12, Fig. 67; Plate 13, Fig. 79).

No. 19

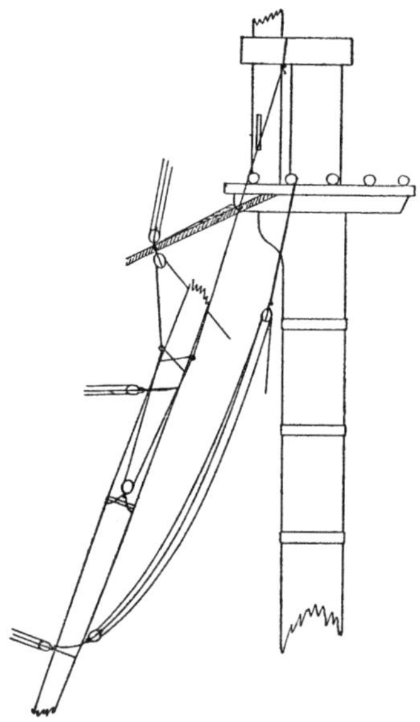

Secure the bends and channels of the ship from chafe by proper means, and heave away, having the *starboard* pendant-tackle in readiness to hook to a strap on the starboard (or *forward*) yard-arm; as also two stout fore-and-aft jiggers to the quarters. When the *after* or port arm rises above the gunwale, and clear of the lower-rigging, cast the yard off from the ship forward, and as it is swept aft by the

*When there is no tide the reverse of this should be observed.

current, hook on the pendant-tackle and jiggers. With these sway it on board, and land it athwartships, on chocks, for that purpose, where it should be lashed to keep it steady; observing the precaution to place a shore under the slings, to prevent it from springing. After which, take off the blocks, bull's-eyes, burton, and tackles, and send the hawser down from aloft.

Now tar the yard-arms, and proceed to place the rigging on in the following routine:

1. Sling-band.—This is a modern substitution for the old D-thimble,* consisting of a stout iron band ,(Plate 13, Fig. 83,) going around the centre of the yard, with an eye-bolt on top, to which the slings are attached by means of a shackle. This falls within the province of the spar-maker, rather than the rigger.†

2. Quarter-blocks‡ (Plate 12, Fig. 68; Plate 13, Fig. 80) are neatest if made of iron, with a stout broad sheeve, (of the same metal, when chain-sheets are used, or of *lignum-vitae,* with rope or hide,) secured to the yard by a band, which is put on by the blacksmith. When fitted with rope, as is frequently the case in the navy, the blocks (which are large single ones§) are seized into *doubled* grommet-straps, (that are simply served, and a strip of leather put on the *after* part, as a preventive from chafe,) with a round lashing; after which, the straps are secured to the yard *on top* by a rose-seizing—the bights, or eyes, being hove close together with a heaver and Spanish windlass. The straps are then spanned towards each other with a lashing underneath, to keep the blocks from spreading.

3. Truss-straps and Pendants (Plate 12, Figs. 77, 78).—The adoption of the *patent iron truss* (Plate 13, Fig. 85) throughout the merchant service, and on board of all government vessels, below frigates, has done away with the necessity of the strap and pendant of "earlier times." But in large class ships-of-war, the "old fashion" (fitted as follows) still prevails. The *straps* (into which the truss-thimbles are seized) are formed of long and short grommets, made like those for the tye-blocks, and lash on each side of the slings with a rose-seizing *on top,* having the thimbles a little above the after centre of the yard. The *pendants* are best when made of four-stranded hide-rope, and leathered on the standing-part, which is made with an eye, and rove around the yard.|| The other end (when rove) is fitted with a single block, the strap of which is formed by the bights of the strands.

4. Clew-garnet-blocks.—Although iron-bound, both in the navy and merchant service, they are sometimes fitted with rope-straps, (Plate 12, Fig. 69; Plate 13, Fig. 81,) which are single, with lashing-eyes, going around the yard, and

*Vide Note for the manner of lashing it on.

†On board of vessels of war there are usually two additional bands—one on each side, for the *preventer* slings

‡In many ships, a small single or double block is secured to the centre of the yard (*prior* to putting on the quarter-blocks) by a rope-strap fitted with lashing-eyes, for the *slab-lines.* (Plate 12, Fig. 72).

§in the merchant service they are more frequently *double* than single, in order to receive the reef-tackle, which is rove like that for the topsail-yard

||One standing-part is placed *outside* of the thimble on the yard, and the other *inside.* This is done to avoid chafe, by crossing them, which otherwise would be inevitable. The cross-jack-yard, however, has but one truss, and consequently a single strap.

secured on top by rose-seizings. They should, also, be wormed and served, and protected from chafe on the after-part by leather (Plate 13, Fig. 42).

5. STRAPS AND THIMBLES, for standing-part of clew-garnet (Plate 13, Fig. 82). These are single, and simply served, with lashing-eyes, which secure on top of the yard by a rose-seizing, *outside* of the clew-garnet-blocks.*

6. ROLLING-TACKLE STRAPS.—Vide Ibid. TOPSAIL-YARDS (Plate 9, Fig. 44).

7. STRAPS AND THIMBLES, for studding-sail boom, (somewhat more than half-way out,) fitted like that for the standing-part of clew-garnet. They are only used on the *fore* yard, except in a few ships, which carry a main-topmast studding-sail.

8. BURTON-STRAPS.—Vide Ibid. TOPSAIL-YARDS (Plate 12, Fig. 70).

9. HEAD-EARING STRAPS.—Vide Ibid. TOPSAIL-YARDS.

10. JACK-STAYS.—Iron ones are universal now throughout the merchant service; and by a recent regulation of the Navy Department the *bending* jack-stays of all government vessels are of iron also But the late method of fitting the courses without after-points, renders *reefing* jack-stays necessary. These are invariably of rope, and fitted in the same manner as those described under the head of TOPSAIL-YARDS, with the exception, that in lieu of being secured by staples, *eye-bolts* are used,† which allows a space between the jack-stay and yard, for the passage of the points in reefing (Plate 12, Fig. 71).

11. FOOT-ROPES.—Vide Ibid. TOPSAIL-YARDS—the only exception being that the inner ends, in lieu of crossing abaft the mast, are connected to each other in the slings by a lashing *underneath* the yard, and this is seized up to the span of the topsail sheet-blocks. In the merchant service they are frequently secured to the arm of the patent truss (Plate 14, Fig. 92).

12. BRACE-BLOCKS.—Vide Ibid. TOPSAIL-YARDS.—Sometimes, in merchant ships, the brace-block is fitted with a long pendant, which connects to the shoulder-band, by means of a hook thimble (Plate 12, Figs. 74 and 76; Plate 13, Fig. 86).

13. LIFTS.—In all heavy ships the lifts are double, and reeve through single blocks, fitted with grommet-straps (which are wormed and served) going over the yard-arm outside, and last of all; the standing-part securing with a hook, to an eye-bolt in an iron crescent extending across the lower cap, (Plate 13, Fig. 84,) while the other part is rove through a single block at the cap, and from thence to the deck, where a small jigger-block is spliced into the end, or attached by means of a hook and thimble. In the merchant service the lifts are single—connecting to the shoulder-band on the yard by means of a hook, and reeving through a block (or bull's-eye) secured to the lower cap (Plate 12, Fig. 75).

Lastly. See that the carpenters and blacksmiths have attached all the necessary appendages to the yard, as cleats, boom-irons, &c., and send the chain-slings up by

*Necessarily so, for otherwise the parts would *cross* and chafe each other.

†Observing, however, before reeving them, to place the upper eye of the stirrups over the bolt, by which a seizing is saved, and the jack-stay made to serve the purpose of a toggle.

one of the top-burtons, fitting them over the lower cap in a score usually cut for that purpose, and securing the two parts by a back-lashing abaft the mast-head. Pass a stout strap around each quarter of the yard, near the sling-band, and hook the lower blocks of the pendant-tackles to them—the upper ones being shifted to a heavy strap over and around the lower cap; hook, also, both top-burtons to the burton-straps on the yard, and reeve the braces. (Vide RUNNING-RIGGING.) Reeve and man the lifts, and sway aloft, keeping the yard clear of the lower-mast by the fore-and-aft tackles. Now when high enough, i. e. when the centre of the yard is a little above the futtock-staffs, key the truss, shackle the slings—unhook the pendant-tackles, burtons, &c., and square the yard by the lifts and braces.

This routine hardly answers for the cross-jack yard (Plate 14, Fig. 87,) there being necessarily many exceptions; among which, we may note the omission of jack-stays, clew-garnet, and slab-line blocks, as also rolling-tackle, head-earing, and burton-straps. It should be remarked, too that the brace-straps, (Plate 14, Fig. 89,) in lieu of going on after the foot-ropes, are secured by a rose-seizing *inside* of the shoulders, about one-sixth of the length of the yard-arm, having the breech of the block to rest against the forward side. This is done to obviate the tendency it has to slip off, arising from the sharp angle which the yard and brace form with each other, when close-hauled (Plate 14, Figs. 90 and 91).

No. 20

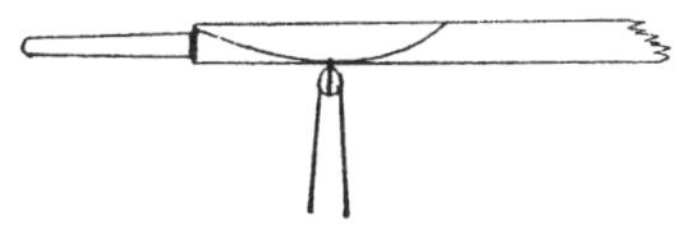

A better way, however, is to fit them with a span, going over the yard-arm, and seizing to a staple on top, as done on board of the frigate "Belle Poule" in 1845. The merit of this plan consists in affording all the advantages of a brace at the yard-arm, without the sacrifice of *leverage* incident to the usual method in bracing up (Plate 14, Fig 88).

NOTE.—When the old D-thimble is used, it is lashed on as follows: Parcel the centre of the yard in the wake of the thimble and lashing, having the parcelling wide enough to turn back on each side, and cover all. Place the thimble on the yard, with the score standing fore and aft; observing to parcel it well. Middle the lashing—usually a small strand of well-stretched rope, and pass the turns with both ends, from the centre out, heaving them well taut with a Spanish windlass, and racking each other in succession; taking the precaution that the turns do not ride each other on the yard. When all are passed, jamb the end between the lashing —haul over the parcelling, and sew it down neatly, covering the after-part with leather.

WHISKERS.

These consist of two gaffs—one on each side of the bowsprit, fitted with jaws, (Plate 15, Fig. 93,) or hooks, (Plate 15, Fig. 94,) like the dolphin-striker; and when compared with the complicated method of conveying support to the head-booms, by means of the old-fashioned spritsail yard, (Plate 15, Fig. 95,) are much simpler, and more frequently preferred; for reason of the little trouble they occasion at sea, by rendering the operation of "canting" unnecessary. This plan has been almost universally adopted in the navy; although in the merchant service, and on board of the smaller class of war vessels, whiskers are seldom if ever met with, except in shape of a stout iron bar (at which the head-guys are secured to eye-bolts) diverging from the bow, at an angle with the cat-head*—a method obviously objectionable, on account of bringing nearly all the strain on the martingale, which is evident, whether we consider the vessel as careening under a pressure of canvas, or regard the jib as a *lifting* sail. The manner of fitting whiskers, however, varies in almost every ship, and depends as much upon the rigger's fancy, and ideas of security, as upon any special rule of seamanship. The following method, adopted on board of the frigate "Cumberland,"† was found to answer every condition; and is given here, for the guidance of those who may be at a loss which plan among the many to employ.

Sway the whiskers on board with a tackle on one of the forward swifters, and land them on the forecastle, with the small ends resting on the head-rail, at each side of the bowsprit. Rack the topmast-stays in the bees (in order not to disturb the position of the mast) before coming up, and reeve the ends through their respective holes in the jaws, taking the precaution to leather them on account of chafe. Then with the assistance of a jigger, from each fore-topmast stay, hooked to a strap around each whisker, (at the distance of about one-third from the outer end,) sway them out, until far enough to rig; and put over the *jib-guys,* which should be fitted with a cut splice, made to set exactly,‡ and so placed as to keep the whiskers perfectly square with each other (Plate 15, Fig. 96). These splices are served, and covered with canvas. Now reeve the flying-jib-guys, (protected by leather at their points of contact,) through the holes bored in the whiskers for that purpose, and prepare to get them into their places—observing to have the bowsprit coppered in the wake of the ends—employing jiggers from the cap, hooked to straps around each jaw. With these rouse them forward, slacking the stay-jiggers, (which in this case, will serve the purpose of lifts, unti they are secure,) and hauling on the jib-guys, in order to bring the whiskers athwartships. Lastly, pass the jaw-lashing, put over the *jumper-brace* and *standing-part of forward lower boom-guy*, then set up the ends of the topmast-stays and the jib, and flying-jib-guys, and take off the jiggers.

NOTE.—It is well to call the attention of the reader to the difference between the jumpers used at *sea* and *in port*—the latter being smaller. They are usually fitted with hook-thimbles in the lower ends, which connect to eye-bolts in the cut-water; and thimbles simply in the upper ones, lashing to the extremities of the whiskers.

*This latter plan coincides in part with the definition of whiskers given by Brady, who makes a distinction, and arranges the first method, under the head of "HALF SPRITSAIL-YARDS"—a classification, though quite independent of the "pent-up Utica" of nautical technology, possessing at least the merit of *originality* with "the author."

†They were substituted for a stationary spritsail-yard, across the knight-heads, after her arrival in the Mediterranean, and were formed out of a spare main-top-gallant-yard fitted with jaws, having holes bored in them to receive the stays and jaw-lashing.

‡A good plan for fitting them is to form the eye by splicing a short piece of the same sized rope, into the guy: should it then be necessary to alter the eyes either forward or aft, on account of stretching, it would only be requisite to draw the ends, and resplice them at the proper place. This method was adopted on board of the "Cumberland," and by keeping the guys on a stretch for three or four days, previous to fitting them, they required altering but *once*, during a period of more than eighteen months.

When preparing for sea, slack the guys, (which must be set up again when the jumpers are secured,) and get the outer ends of the whiskers down to an angle of about 20° with the horizon. Then hook the (stout) jumpers to the cut-water, and set them well taut with a laniard, which is spliced into the thimbles of the upper ends.

It may not be altogether amiss here to add a "wrinkle" in *squaring* the whiskers in *port;* which may be done by simply clapping on the *guys,* and slacking the jumpers—*without* the necessity of a (troublesome) "*toppin*'-rope," through tail-blocks on the fore-stays, as *sometimes* practised by newly-fledged boatswains, when first "called" in the service.

TOP-GALLANT AND ROYAL YARDS

Having brought the yard (Plate 16, Figs. 97, 106; Plate 17, Fig. 111,) (fitted with an iron sling-band and jack-stay*), (Plate 16, Fig. 98) alongside, sway it on board with the long yard-rope, rove through the sheeve-hole in the top-gallant mast-head, by bending it on to the slings, and stopping it down to the forward yard-arm. Place it in the gangway on chocks—unbend the yard-rope, and prepare to place the rigging on, in the following order:

1. PARREL (Plate 16, Figs. 102, 108).—This is best when formed of single grommet-straps (a long and short one) going over the yard, on each side of slings, at a distance apart equal to the diameter of the top-gallant-mast, and cleated, to prevent slipping. These are made to fit snugly, by means of round seizings passed close to the yard on the after-part, leaving a surplus bight to each, which forms the lashing eyes—that of the bigger strap (marled together and leathered) being long enough to go around the mast, and meet the short one, (leathered also,) where it is secured by a lashing of small-sized ratling stuff, after the yard is crossed. In the merchant service, and on board of the store-ships, *jaws* are frequently put on the top-gallant and royal yards, in lieu of parrels, and have been found to answer every purpose; there being, however, a necessity for care in sending them up or down, lest the curve or shoulder of the jaw should be knocked off (Plate 17, Fig. 112).

2. QUARTER-BLOCKS (Plate 16, Figs. 99, 107).—These are double, and invariably iron-strapped, connecting with a hook to a band on the yard. This is by far the best plan, as it saves a deal of trouble in sending the yard up or down. When rope is used, the blocks are single-strapped, with spliced eyes going around the quarters, and secured by rose-seizings on top.

3. STRAP AND THIMBLE, or *becket,* for the standing-part of the clew-line. Vide Ibid. TOPSAIL-YARDS (Plate 16, Fig. 100).

*In the "Cumberland," on the top-gallant and royal-yards, the jack-stays spliced into the eyes of the foot-ropes, around which, on top, cringles were worked for the head-earings; a plan answering all the conditions of security, and in point of neatness, worthy of universal imitation.

4. STRAP AND THIMBLE for rolling-*rope* (Plate 16, Fig. 101). (Vide Ibid. TOPSAIL-YARDS). This is omitted by many seamen, who deem it unnecessary, and send the yard down, when the sail is taken in. But it would be well to remember, that if it be kept aloft, and the rolling-ropes passed securely over the topmast cap, a considerable degree of support might be given to the top-gallant-mast, as the yard-arms thus act the part of out-riggers, and the lifts perform the service of shrouds.

5. STRAP AND THIMBLE, for head-earing. Vide Ibid. TOPSAIL-YARDS.

6. FOOT-ROPES (Plate 16, Figs. 103, 109 and Plate 17, Fig. 113). Vide Ibid. TOPSAIL-YARDS. The inner ends, however, do not cross abaft the mast after the yard is sent up, but seize to the quarters.

7. LIFT (Plate 16, Figs. 104, 110 and Plate 17, Fig. 113).—This is single with a round-eye (the splice of which is served) going over the yard-arm, (when swayed up for crossing,) and generally leathered to make it slip readily, off or on, as the case may be. In the naval service, it is cut of sufficient length to reach the top, after reeving through the bull's-eyes, between the top-gallant shrouds, and marled to the brace when the latter is single, in order to hold on, or come off together. But in merchant ships of the present day, the lifts are most always made to go *standing*, and secured to a becket, or thimble, under the eyes of the top-gallant rigging; a plan which is quite at "loggerheads" with square yards, and productive of "high-sounding" anathemas from the men in getting them over the yard-arm, particularly in a heavy swell, on the "threadbare" poles of a down-east merchant-man.

8. BRACES (Plate 16, Fig. 104 and Plate 17, Fig. 113).—When fitted double (which is nearly always the case in the merchant service) a single block is strapped into a grommet, which is wormed and leathered, and sets snugly over the yard-arm when swayed aloft. In this way it is not unfrequently marled to the eye of the lift.

Lastly. Splice the *snorters* (which are cut about half the length of the yard-arm, and simply whipped at the end) into the thimble of an eye-bolt at the yard-arm, hitching it to the rolling-rope strap; and prepare for crossing. The yard-rope, having a lizard attached, (overhauled down forward, and outside of all,) is rove through a good-sized grommet passed over the (upper) yard-arm, and hooked to the bolt in the sling-band; the lizard being rove through the rolling-rope thimble, and hitched to the one on the opposite quarter. Take the eye of the lower lift, and brace in the topmast rigging, and that of the upper one to the opposite side of the cross-trees, and sway aloft. When the upper yard-arm rises above the heel of the top-gallant-mast, take off the grommet—slip on the lifts and braces, gathering up the slack of the lower one, and sway away, until the slings of the yard are over the topmast cap; then take a turn with the parrel-lashing abaft the mast, through the eye in the opposite strap—tend the lifts and braces, jerk on the lizard, and sway across, squaring the yard, and passing the parrel for good and all.

ROYAL YARDS.

The routine of rigging and crossing is precisely similar to that of the top-gallant-yards—the only difference being that the quarter-blocks are single, and straps for the rolling-rope and standing-part of clew-line are omitted (Plate 16, Figs. 105, 106, 107, 108, 109 and 110; Plate 17, Figs. 114 and 115).

TRYSAIL-MASTS—GAFF AND BOOMS.

These spars are usually swayed on board by means of the fore or main yard and stay-tackles—purchases most frequently in use, and convenient at this stage of the equipment.

TRYSAIL-MASTS.

Of late an iron railway, (Plate 18, Fig. 116,) on the after-part of the lower-mast, with sliding-bolts, to which the sail is bent, has been substituted for the trysail-mast of a less recent date, particularly in the vessels of the naval service,—a plan which possesses many advantages in point of looks, but attended with an infinite degree of trouble on account of the tendency in the bolts to slue and jamb, when hoisting or lowering the sail. In the merchant service, however, and on board of the store-ships, this method is not adopted; the gaff being frequently stationary with hanks on it, for the sail, which is bent to an iron jack-stay, (Plate 18, Fig. 117,) secured by staples, or to a wooden batten (Plate 18, Fig. 118) on the lower-mast, and set, or taken in with *in-and-out haulers;* although a preference exists in many instances for the old trysail-mast, which is shipped by means of a tackle hooked to a strap above the futtock-band; the head being pointed through a hole in the after-chock, and the heel (over which the hands are passed) stepped in a socket, or mortice, on the fife-rail, or a shoulder, bolted to the lower-mast just above the deck; after which, the head is secured by a lashing through a B-cleet underneath the top, or with iron keys; copper having been put on, in the wake of the gaff (Plate 18, Fig. 119).

NOTE.—In merchant vessels, the *fore*-trysail is sometimes dispensed with, and its absence supplied by the main-staysail; but it is nevertheless worthy of remark, that the Americans are almost the only ones who equip their ships with a trysail, forward; an exception being sometimes met with among the Danes and Russians. This "mark of distinction," or the absence of it, is often serviceable in *guessing* a vessel's nationality.

GAFFS.

These are fitted in a variety of ways,—having jaws, with a trysail-mast; stationary, and secured to the lower-mast by means of eye-bolts within each other, (like lock-thimbles,) when a batten or jack-stay is used; or with a sliding-chock, when fitted "railway fashion." The introduction of iron has done away with much of the gear incident to gaffs in former times; and the *vangs* and halliards are about all now that fall within the province of the rigger—the brail-blocks being superseded by sheeves (for that purpose) in cheeks fitted by the carpenters on each side of the gaff (Plate 18, Fig. 120); while the blocks for the throat and peak halliards are iron-strapped, and connect to bands with eye-bolts.* (Plate 18, Fig. 121). The vangs are best when fitted with a cuckhold's-neck, which is wormed and covered with canvas, having a neat seizing at each side, and cleated both forward and aft, to prevent slipping. The gaff is sent up by the halliards, and the jaw-lashing passed, or the eye-bolts riveted, as the case may be.

NOTE.—The *spanker*-gaff (Plate 18, Fig. 123) should always be fitted with throat and peak halliards to hoist and lower, as necessary; for otherwise it would be almost impossible to reef the sail, which by-the-by is frequently the case in the merchant service. In brigantines and schooners, it is not unfrequently the case that eye-bolts (Plate 18, Fig. 122) are attached to each side of the jaws, for preventer lashings in heavy weather; and a single block (grommet-strapped) is put over the gaff-end for a down-haul; vangs being dispensed with as useless, on account of the sharp angle at which they act, in consequence of the height of the gaff.

BOOMS.

That for the *spanker* (Plate 19, Fig. 126) is neatest, if shipped with a goose-neck to an eye-bolt on the mizen-mast, and fitted with an iron band (Plate 19, Fig. 124) over the boom-end for the topping-lift, (when single,) and the guys, both of which connect to it with sister-hooks (Plate 19, Fig. 125). The *sheet-blocks* are best if strapped with rope-grommets, on account of the jerks and checks of gibing, which render eye-bolts liable to snap and break at the neck. The *foot-ropes* are formed with round-eyes (wormed and served) going over the boom-end, and seizing to staples inside of the taffrail. The topping-lifts (one on each side) are usually fitted with sister-hooks in the ends to an iron span, about one-fifth of the extreme length of the boom from the outer end; while the running-parts reeve through blocks at each side of the mizen lower tressle-trees, and thence to the deck, where gun-tackle purchases are attached. In men-of-war, particularly those of the smaller class, and in the merchant service, the topping-lift is not unfrequently single, and rove through the gaff-end, and a roller in the after-part of the mizen lower-cap, having a jigger connected to it, by means of a hook and thimble. The *sheets and guys* are best if rove in *one*, which may be done, by either having the standing-part over the boom-end, or attached to the blocks at the quarters. (Vide Ibid. RUNNING-RIGGING.) The *out-haul* (going around the boom with an eye, is rove through a block on the clew of the sail, and a sheeve in the boom) should be of *hide*-rope, which is far better calculated than hemp to stand the wear and tear, and friction of constant use. On the *main-boom* of brigantines and schooners, (Plate 19, Fig. 127,) the topping-lift is usually fitted with the standing-part secured at the mainmast-head by a hook in an eye-bolt of the wythe; while the lower end is spliced around a double block, in which a fall is rove, leading through a single one, and a sheeve in the boom. In this class of vessels, no out-haul is used; the

*In the "Fairfield," where the trysails were fitted with a railway, the gaffs were stationary, and supported by a single leg, (from a bolt in the after-chock,) spliced around a thimble on a span attached to the gaff,—a method which, though neat, was nevertheless troublesome in holding the sail.

clew of the sail shackling to a band around the boom, inside of which an iron traveller, covered with leather, is shipped for the purpose of securing the reef-cringles in reducing sail, each successive reef bringing the traveller nearer to the mast. A heavy strap, (which is cleated forward,) with thimbles at each side, is put around the boom about halfway out from the jaws, for the boom-tackle pendant, which is fitted with a hook and thimble in each end, and used only in going large. The *crotch-ropes* (an essential requisite in the management of a boom-mainsail) are usually served all the way, and fitted with a hook in one end for a stout eye-bolt in the quarters—the boom being protected by copper in the wake of them, just forward of the sheet-blocks, which should be double-strapped with rope. These, with the addition of belaying-cleats for the topping-lift and reef-pendant, (fitted of hide-rope, and reeving through a sheeve in the boom, with a jigger attached,) complete the equipments. The boom is then got in its place by means of the *throat* halliards and topping-lift, assisted by guys and athwartship tackles, as requisite.

STUDDING-SAIL BOOMS.

That for the *lower* studding-sail (Plate 20, Fig. 133) is fitted with an iron goose-neck and key, which connects to a bolt in the forward part of the fore channels; and is shipped either by means of the fore and main yard-tackles, or with burtons on the standing-backstay and forward swifter of the fore-rigging. On the outer end, about two-thirds from the goose-neck, an iron band is put over the boom, (Plate 20, Fig. 135,) having eye-bolts on the forward, upper, and after sides, for the topping-lift, and the guy-blocks;* and in vessels of war, *mooring* pendants (Plate 20, Fig. 134) with huge thimbles in the lower ends for the boats, and a Jacob's ladder, are attached, by lashings like those used in securing the foot-ropes on the topsail-yard. These, with the *topping-lift*, (Plate 20, Fig. 132,) (which is fitted with sister-hooks, connecting to the boom, and reeving through a small clump block, seized between the forward (lower) swifters, above the futtock-staff, with a double or single purchase in the end for the fall,) and the *out-haul block*, (Plate 20, Fig. 136,) double-strapped with rope, having an eye to set taut over the boom-end, outside all the *standing* gear, falling within the province of the rigger, except perhaps the *lizard*, the thimble of which traverses on the topping-lift, and is used as a fair-leader at the outer quarter of the fore-yard in getting the boom out or in (Plate 37, Fig. 189).

NOTE.—When the boom is not in use, it is hauled alongside by the after-guy, and rests in cranes, shipped for the purpose in the waist—the topping-lift being stopped down the forward swifter. Not unfrequently in English vessels, particularly men-of-war of a large class, the topping-lift is rove through a stationary bull's-eye on the fore-yard, and a span-block at the lower cap—a plan which dispenses with the lizard, but less economical than the method given above, on account of requiring more rope.

The frequency of accident to *topmast studding-sail booms*, (Plate 20, Fig. 131,) at sea, often renders "shifting" a necessary operation, and this consideration is deemed sufficient, by many men of experience, to recommend the fitting of both extremities alike, viz. tapering with a score and hole for the heel-lashing—that in the outer end (while in use as such) being "plugged" with wood, which can be

*In the merchant service, the guys are most frequently single, and are fitted simply with sister-hooks.

readily driven out in case of a necessity for a "shift."* Through the hole in the inner extremity, reeve the *head-rope*—turn a Walker's knot in one end, and whip the other—unclamp the irons on the yard, and prepare for sending the booms aloft. Hook one of the top-burtons to the lower lift, somewhat more than halfway down, and send the fall to the deck "before all," attaching the lower block to a bight formed by hitching the heel-rope over the boom, and stop the standing-part down towards the outer end. Bend a hauling-line, from the slings of the yard, on to the heel, and sway aloft, pulling up steadily on both, until high enough. Then pass slip-ropes over it—point it fair through the pacific-iron, and rouse on the burton, cutting the stops as requisite, until sufficiently far to place the *lower studding-sail halliard-block*, and that for the *tack*, over, fitted with straps setting tautly around the boom-end, and kept from slipping off by an iron pin passed outside of both. After which, clamp the boom-irons; pass the heel-lashing, (through its thimbles on the jack-stay;) rig out to the square mark, and take off the burton, tripping-line, and slip-ropes. Lastly, seize a hook horizontally on the yard at the "rig-out mark," with the point out-board, for the purpose of securing the boom, when setting the sail,† and fit the in-an-out jigger, ready for use.

NOTE.—In many merchant vessels these booms are fitted to traverse *under* the yard. But the plan is objected to, by "many men of many minds," on account of the chafe induced by a constant contact with the topsail sheets, the unhandiness of securing the heel, and the clumsy appearance given to the yards in port—the sole advantage being, that the boom is out of the way in reefing or furling sails.

The *top-gallant studding-sail booms* (Plate 20, Fig. 128; Plate 37, Fig. 189) are rigged nearly in the same manner, with the exception of a halliard-block at the outer end, and a score in the heel; and in lieu of being retained on the yard by boom-irons, it is usually (and best) made to traverse on a rope jack-stay, which is seized to the eye of the topsail-lift, and sets up to its opposite in the slings of the yard. The *tricing-line*, (leading through a small block hooked to a thimble seized between the forward-swifters, close up to the eyes of the topmast rigging,) when required to rig the boom out, is made to answer the purpose of an in-and-out jigger, thus: The block is hooked to a stationary thimble well out on the yard, and from thence the tricing-line leads into the top. The boom (fitted with a becket in the heel) (Plate 20, Fig. 129) then toggles to a travelling thimble on the jack-stay; and when required for setting the sail, is secured by means of a lashing passed over a hook on the yard, like that for the topmast studding-sail boom, already mentioned. In the merchant service, this "extravagance" is not common, although it has many advantages to recommend it; but one must bear in mind that "*owners*" and "*captains*" are not *always* synonymous terms.

NOTE.—The booms on the topsail-yard are usually sent up by the halliards, rove through the jewel-block, (Plate 20, Fig. 130,) which is secured to the forward-swifter of the topmast-rigging, and the boom slung in a span. For the manner of reeving the gear, vide RUNNING-RIGGING.

*Should this plan of fitting the boom be preferred, and it becomes necessary to shift it to the opposite yard-arm, for use, it may be done as follows: Unclamp the boom-irons, and rig in, until clear of the "pacific," taking off the blocks at the end, and changing them to the other side, ready for placing over when the boom is transferred. Trice the heel up to the lower-stay collar, (securing it there with the lashing,) and lower the outer end down by means of the topmast studding-sail halliards, rousing it up on the opposite yard by the other halliards, overhauled down outside of all for that purpose. Then slack the heel-lashing, rig out, put over the blocks, clamp the irons, &c.

†The heel-lashing is passed over the hook, and back through the score in the boom, and two half-hitches taken with the end around all parts.

RUNNING-RIGGING.

In the fitting of running gear there are two essential observances—economy, and an assimilation in size between the ropes and blocks, both of which may be attained by following the rule laid down by Glascock, who says it "is not recommended to *cut* and reeve, but, on the contrary, to *reeve* and cut"*—a principle which is founded on the "fundamental formula" of *all* professions—viz. *common sense*. Having stretched the rigging, (by "thorough footing" it, i. e. passing the end through the coil, and then taking it to the capstan,) commence reeving it in the order of its importance, in which we may "reckon," as first,

BRACES.

(Plate 22, Fig. 140, and Plate 24, Fig. 145)

FORE-BRACES (Plate 21, Fig. 137).—In the navy, the standing-parts are rove through holes in the fore part of the main-bibbs, (having overhand knots in the ends;) through the blocks on the yard, (from *in, out,*) thence back through others on the outside of the bibbs, and down to the sheeves in the fife-rail. (usually) from *aft, forward.* In merchant vessels, (Plate 23, Fig. 144,) the standing-parts of the fore-braces are most frequently made fast to the strap of a single block in each of the yard-pendants,† and thence rove to double ones on the bulwarks just forward of the gangways, having the hauling parts in the second sheeves of the latter. This plan requires less length, and yields a better purchase than the method practised in the navy. But an occasional unkindly "clip" from the clew of the mainsail in tacking, and a constant exposure to dampness and wet, often illustrates the fallacy of this *economical* rule.

MAIN-BRACES (Plate 21, Fig. 138).—The standing-parts are clinched, or fitted with running-eyes, over the bumpkins aft—then rove through the yard-blocks from *out, in,* back to others (*inside* of the standing parts,) and through sheeves in the bulwarks. In the merchant service the main braces are rove like the fore (Plate 23, Fig. 141).

CROSS-JACK BRACES (Plate 21, Fig. 139).—The standing-parts are spliced into the strap of a *double* block‡ hooked to an eye-bolt on each side of the mainmast, in a line with the yard§—thence to the yard-blocks from *out, in,* and back to the inner sheeves of those on the mainmast.

*If the absence of quotation marks, and the omission of italics, be any criterion of *originality*, it is certain that "the author" of the *Kedge Anchor* must have been afflicted with some pretty strong *symptoms* of this very same excellent idea, when *he* "recommended, *not* to cut and reeve. but, on the contrary, to reeve and cut;" particularly, if we analyze, somewhat *critically* the expression of it For example: "the contrary" of "*not* to cut and reeve," is evidently "to cut and reeve;" the coincidence with Glascock being "passing strange," except perhaps, that he wrote with "a pen," and Brady "with the marlinspike," with which he acknowledges himself so well "versed;" a circumstance which may account for the *negative* getting *astern* of the *past participle*. (Vide Preface, and page 249 of the latter's "work.")

†Vide BRACE-BLOCKS, LOWER-YARDS.

‡The outer sheeve of which then answers for the mizen top-bowline.

§The reason is obvious: for if it were otherwise, the angular action of the brace would cant the yard either up or down, and consequently slack one or the other of the mizen-topsail leeches.

FORE-TOPSAIL BRACES.—The standing-parts are hitched above the eyes of the main-topmast rigging, and seized to the collar of the stay, (to avoid chafe from the foot of the top-gallant-sail;) thence forward, and *down* through the yard-blocks to others on the main-stay, and at the bibbs. In merchant ships, the yard-blocks are generally attached to pendants, and the braces, in lieu of leading to the main-stays, are brought to the bulwarks like those for the lower yards.

MAIN-TOPSAIL BRACES (Plate 23, Fig. 142).—The standing-parts are hitched to the mizen-topmast-head, and seized to the stay-collar; thence rove to the yard, and down to span-blocks on the mizenmast, about halfway between the top and deck. In many ships the standing-parts are secured to a strap which travels up and down the mizen-topmast with the parrel, (in order to shift the strain lower down, as it becomes greater, and sail is reduced,) a plan frequently met with in the merchant service, and possessing many commendable features.

MIZEN-TOPSAIL BRACES (Plate 25, Fig. 146).—The standing-parts are secured to the strap of a block at each side of the main lower cap; thence to the yard, and back to the blocks, and so down through the lubber's-hole to the deck.

FORE-TOP-GALLANT BRACES.—In the navy they are usually rove single, the standing-parts going over the yard-arms with the lifts—thence through leading blocks on the main topmast-stay collar, and others, under the eyes of the topmast rigging —whips (the standing-parts of which are secured to the lower tressle-trees) being attached to the ends. In merchant-men the top-gallant braces (Plate 23, Fig. 143) are invariably rove double, through blocks going over the yard-arms, and the standing-parts seizing to the topmast-stay collar; this plan renders whips unnecessary, the purchase being deemed sufficient without them.

MAIN-TOP-GALLANT BRACES.—Vide preceding, and substitute *mizen* for "main."

MIZEN-TOP-GALLANT BRACES (Plate 25, Fig. 147).—In reeving these braces, which are almost always single, there appears to be but one method, viz. through the inner sheeves of small double blocks, underneath the main-topmast cross-trees—the outer sheeves being used for the bowlines.

FORE-ROYAL BRACES.—These are single, (without whips,) and rove like the top-gallant braces, except that they are taken to the main-top-gallant mast-head and stay, in lieu of lower down.

MAIN-ROYAL BRACES.—Same as fore-royal braces, except taken to the *mizen*-top-gallant mast-head, and through blocks on the stays.

MIZEN-ROYAL BRACES.—Single, over the yard-arms, and through middle-sheeves in the after-chock of the main-topmast cross-trees, the outer ones being used for the royal bowlines.

TOPMAST STUDDING-SAIL-ROOM BRACES.—These (which are confined chiefly to vessels of war) may be either single, going over the boom-end with a running-eye, and leading through a tail-block on the forward-swifter of the main rigging; or double, with a pendant and whip leading to the main-rigging also.

PREVENTER BRACES (Plate 25, Fig. 148).—These are fitted variously, but the best and most economical way, is with a *pendant* and *whip*, the former going round the yard, hooking to its own part, and the latter led to the deck, well aft.

HALLIARDS.

TOPSAIL HALLIARDS (Plate 27, Fig. 151).—Where double-tyes and gins are used, the standing-part of the halliards is made fast to a single block, (which is iron-strapped and fitted with a swivel,) in each side of the channels, and then rove in a double one connected to a thimble in the end of its respective tye, by sister-hooks. In the merchant service, (Plate 27, Fig. 152,) and on board of the smaller class of war vessels, where gins are dispensed with, and a single-tye is employed, a runner is rove through a single block spliced into the end of the latter, and the double one of the halliard purchase* hooked to an extremity of the runner, which is fitted with a thimble; the other (or standing-part) making fast to a bolt in the after-chock. When the halliards are used, the hoisting part is rove through a leader on deck.

TOP-GALLANT HALLIARDS (Plate 26, Fig. 149).—On board of large class ships of war, it is usual to fit the top-gallant-yard with a "regular" tye and runner. But this plan is manifestly too expensive and troublesome to meet with much encouragement in the merchant service—a yard-rope being also requisite, in case of sending the yard up or down. Indeed, the best and most economical method is to have a double-and-single tackle hooked to a bolt in the lower tressle-tree, and fitted with two lizard-tails, which connect to the yard-rope, just below the eyes of the topmast rigging, the hauling part of the purchase being sent through the lubber's-hole to the deck, and the surplus end of the yard-rope coiled, or flemished down snugly in the top.

ROYAL HALLIARDS (Plate 26, Fig. 150).—These are best, if fitted with a simple whip, which is formed thus: The yard-rope being rove in a leader on deck, is passed through a single block, (fitted with a strap having an eye,) and toggled on at some distance above the bulwarks, as represented in the annexed figure. In the event, then, of having to send the yard down, it is only necessary to take off the block, which will leave the yard-rope clear for running.

No. 21

THROAT HALLIARDS (Plate 28, Fig. 153).—If for a spanker, or trysail, they usually consist of a purchase rove through double and single blocks; the former hooked to a bolt on the under side of the after lower-chock, and the latter to a band and eye-bolt on the inner end of the gaff; the hauling

*This block is usually connected by means of spider-legs to a *clamp-ring* (which is parcelled and leathered) around, and travelling on the standing-backstay, in order to keep the turns out of the halliards when let go, and the block from chafe by contact with the top, (and the *hard* heads of "curious people" lounging over the rail.)

part leading through the upper block from *aft, forward,* to the deck. In brigantines, and vessels with a boom-mainsail, both blocks are double.

PEAK HALLIARDS (Plate 28, Fig. 153).—The best plan for peak halliards is to reeve them as follows: Splice the standing-part into the breech of the mast-head block, (which is double,) and reeve from thence, through the *inner* block on the gaff, from *aft, forward;* then up through the *port* sheeve of the mast-head block—out through the block at the gaff-end, from *forward, aft;* and lastly, back to the starboard sheeve of the mast-head block, as illustrated in the accompanying cut.

No. 22

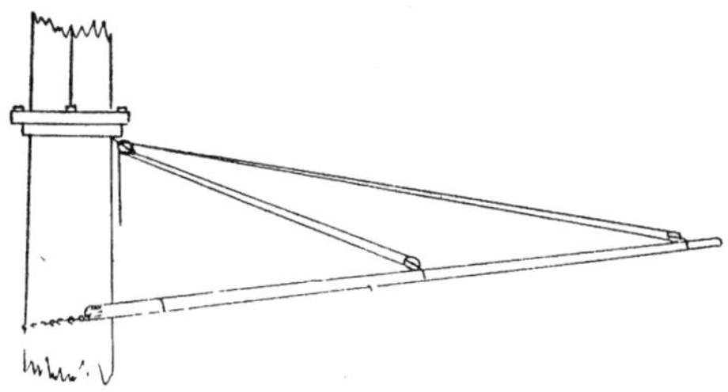

The merit of this system will be apparent, if we consider that the hauling part, by being rove *last*, at the gaff-end, permits the peak to drop the instant the halliards are let go.

STORM-STAYSAIL HALLIARDS (Plate 28, Fig. 155).—The following simple method of fitting the stay and halliards *in one,* is recommended, as obviating in a great

No. 23

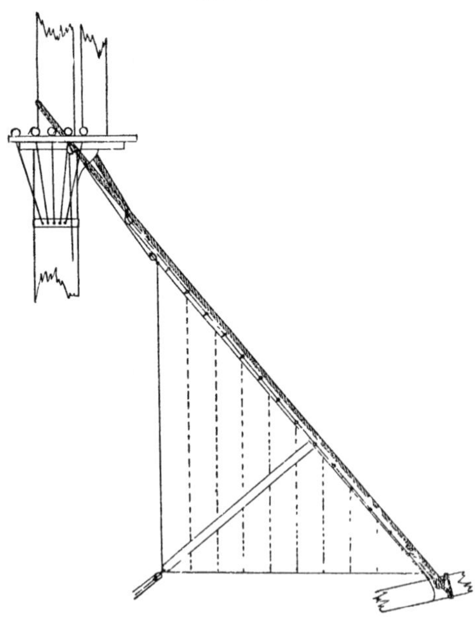

measure the expenditure of rope for "staysail gear," so universal both in the navy and merchant service. The *stay,* which is somewhat longer than the luff of the sail, and of the necessary size, is rove through a stout thimble seized into a strap around the bowsprit, (or lower-mast, as the case may be,) and turned in cutter-stay fashion, (Plate 28, Fig. 154,) or connected by means of a hook. The *halliards* being rove from a header under the top, to a block on the head of the sail, and another at the crotch of the lower-stay collar, (as in the preceding figure,) are then spliced into the upper end of the stay, which is passed through the hanks. The advantages of this plan consist in tautening the stay as the sail is hoisted, and in dispensing with a downhaul, which is rendered unnecessary, by letting go the halliards, and hauling aft on the sheet.

JIB AND TOPMAST-STAYSAIL HALLIARDS.—The best manner of leading these ropes is a matter of opinion, formed, in a great measure, from whatever may be uppermost in the mind of the rigger; as fancy, security, neatness, or desire for "professional greatness" in the shape of a new invention—contingencies often productive of some *grand faux pas* in subsequent practice.* In the store-ship "Southampton," while attached to the African squadron, the jib and staysail halliards (Plate 29, Fig. 157) were rove in the lower sheeves of iron fiddle-blocks, hooked to a bolt in each side of the topmast tressle-trees—the former double, through a block on the head of the sail, with the standing-part lashed to the crotch of the stay-collar; and the latter single, with sister-hooks to the head-cringle, and a whip, the block of which came just below the eyes of the rigging, when the sail was taken in. In point of neatness and security, this plan was found to correspond with all the essential conditions of good seamanship; for, in shifting topmasts, or in performing any other necessary evolution, "touching" these ropes, the blocks only required to be unhooked, while the upper sheeves served for the topsail buntlines.

NOTE.—The jib-halliards (Plate 29, Fig. 156) should be led on the *starboard* side, and those for the staysail vice versa—a rule which is self-evident, when we remember that the latter is set on the *port* topmast-stay. The method of fitting these halliards with *whips*, is not approved of by seamen generally, on account of the liability to tangle and get foul in hauling down the sail; and the obvious necessity of separating the parts widely from each other. And again, where the steeve of the bowsprit is not very great, and the mast is made to rake *aft*, not unfrequently a *certain* mathematical truth about the *hypotenuse* being the *greatest* side of the *triangle*, is comprehended, for the first time, after the aforesaid "whip" has been *cut* from the coil.

FLYING-JIB HALLIARDS (Plate 29, Fig. 158).—These are invariably rove single, through a neat block secured at the hounds of the top-gallant-mast, (on the *port* side,†) under the eyes of the rigging, and connected to the head-cringle on the sail by means of sister-hooks. In large ships, however, they are sometimes rove double, and the standing-part seized to the splice of the stay on the *under* side.

GAFF-TOPSAIL HALLIARDS (Plate 29, Figs. 159, 161).—These are single, and in barks and ships are rove through a sheeve in the top-gallant mast-head, and attached to the yard with a *fisherman's bend* (Plate 29, Fig. 160); or if the sail is triangular in shape, to a becket in the head-cringle, with a *sheet-bend* (Plate 29, Fig. 162). On board of schooners and hermaphrodite brigs, they are rove through a sheeve in the *topmast*-head.

*For example: Fordyce recommends "a very *neat* and handy mode of fitting the jib-stay and halliards," which consists of "a spider-hoop, screwed on to the mast-*head, just under* the topmast *cap*," &c. This "good plan" might have done very well before top-gallant sails came into fashion; but it is pretty certain now, that the *foot* of the sail, and this "stuck-up" stay, would "*wail* and gnash" each other amazingly.

†A block precisely similar to this one is lashed on the starboard side for the top-gallant buntline.

LOWER STUDDING-SAIL HALLIARDS (Plate 37, Fig. 189).—In the navy the *outer* ones are rove through a block hooked into the thimble of the burton-pendant; thence to the halliard-block at the end of the topmast studding-sail-boom, and attached to the yard with a fisherman's bend. In merchant vessels, where the burton-pendants are dispensed with, they are rove through a span-block on each side, which is secured with lashing-eyes *above* the topmast rigging, and forward of the shrouds—the hoisting-part being on deck through the cross-trees and the lubber's-hole. The *inner* halliards are usually formed out of the fore clew-jigger, hooked to the inner head-cringle of the sail, and a strap around the cap-shore, clear of the top rim.

TOPMAST STUDDING-SAIL HALLIARDS (Plate 37, Fig. 189).—These are rove on each side through a single block—the span of which is passed over (and seized together under) the topmast cap, between the doublings of the mast, being sufficiently long to reach forward of the top-gallant shrouds; thence *abaft** the topsail-yard, through the jewel-block, and so to the deck, where they are attached to the central part of the studding-sail-yard with a fisherman's bend.

TOP-GALLANT STUDDING-SAIL HALLIARDS (Plate 37, Fig. 189).—These are rove on each side, through a single block, (which is fitted with a rope-strap and tail,) hitched above the eyes of the top-gallant rigging; thence *abaft*, to the jewel-block, and so to the top, where they are bent to the studding-sail-yard, in the same manner as the halliards previously mentioned—the hoisting-part being sent down *between* the horns of the cross-trees, to the deck abaft, and clear of all.

SHEETS.

FORE AND MAIN SHEETS (Plate 30, Fig. 170).—The standing-parts are connected to eye-bolts on the outside of the bulwarks with sister-hooks, just *forward*† of the sheeves for the hauling-parts; from thence, they are rove *up* through the blocks at the clews‡ of the sail, and back, in-board, through the bulwark sheeves. In the

*Seamen are still at issue on this point, and Brady says "the halliards should be rove on the *fore* side of the topsail yard"—an opinion coincident with that of many others, who argue that the sail, by bellying forward, slues the jewel-block, which brings the halliards (if rove *abaft*) in contact with the shoulder of the yard. This is true to a certain extent, and so long as the yard remains *square*, (which is far from being the case *always* with studding-sails set;) but when the parallelism subsisting between it and the halliards is destroyed by *bracing forward*, the contact and consequently the only *possible* cause of chafe, is removed. If, on the contrary, the halliards are rove forward of the yard, it is evident that they would not only chafe against the clew of the top-gallant-sail, but over the boom on the topsail-yard, and with an intensity and force proportioned to the angle formed by bracing.

†The difference between three feet *forward* of the sheeve-hole, and three feet *abaft* it, is equal to a fathom of rope.

‡These are best, if secured to the cringle by means of a *shackle*, which should be made large enough to receive the tack, and clew-garnet-blocks also. This plan was alopted in 1845, on board of the "Cumberland'—a ship, at that time, distinguished for neatness and efficiency; and it cannot be too strongly recommended, on account of the facility it affords in shifting courses, by a riddance of the clew-blocks, which, as permanent fixtures, render the reeving of tacks, sheets, and clew-garnets, an essential and troublesome feature—to say nothing of the increased weight and room of the spare sail, or the expenditure of blocks.

merchant service, where the tacks and sheets are often single, the clew-blocks are dispensed with, and the standing-parts are shackled, or simply clinched to the cringles.

TOPSAIL SHEETS.—When double, (Plate 30, Fig. 165,) as on board of large class ships-of-war, the standing-parts are fitted with running-eyes, (Plate 30, Fig. 168,) which go around the yard-arms *outside* of all, and thence rove from *out, in,* through the clew-blocks to the yard-sheeves, and the quarter-blocks in the slings; being led, lastly, to the bitts on deck, forward of the mast. If single, (Plate 30, Fig. 166,) (in which case the *reeving* parts should be of *hide*-rope,) they are simply secured to the clew-cringle with *inside* clinches; but where chain is used, as in the merchant service, they are connected by small stout iron shackles. On board of *polaccas,* (a class of vessels peculiar to the Mediterranean,) topsail-sheets are not used, and the clews of the sails are lashed to large beckets on the lower yards, fitted for that purpose.

TOP-GALLANT AND ROYAL SHEETS (Plate 30, Fig. 163).—These are always single, and attached to the clews of their respective sails by sister-hooks. The former are rove through the topsail-yard, to the *after*-sheeves* of the quarter-blocks, (in the wake of which they were best, if formed of hide-rope;) from thence they are led through the lubber's-hole to the deck. The latter are rove in the same way, except through the sheeves and quarter-blocks of the top-gallant-yards, and thence through thimbles on the futtock-staffs of the topmast-rigging (abreast of the *second* shrouds) to the top, or deck, as may be preferred.

STORM-STAYSAIL SHEETS (Plate 30, Fig. 164).—These are temporary purchases, and consist usually of stout deck, or pendant-tackles, hooked (and moused) (Plate 30, Fig. 167) to the clew-cringles, and bowsprit well aft, in order to form, as near as possible, a line with the foot of the sail. The hauling-part should then lead from the *forward* block, by which a greater purchase is obtained; although the reverse of this is advocated by many seamen, on account of the difficulty sometimes experienced in getting a turn with the belaying-end, in consequence of the flapping of the sail; but this objection will be entirely overcome, if the sheet be hauled aft, and the foot taut, before hoisting.

TRYSAIL SHEETS (Plate 30, Fig. 169).—The best plan for fitting these is to have a long pendant attached to the clew of the sail for the fall to hook into, as it saves the trouble of "lighting up" the blocks to hook and unhook in shifting the sheet, as in wearing ship, &c.

JIB AND TOPMAST-STAYSAIL SHEETS (Plate 31, Fig. 173).—Both of these are generally fitted with a whip and pendant, as follows: The latter, which should be in length equal to about one-third of the foot of its respective sail, is wormed and served; and after being middled, a good-sized thimble (around which the parts of the pendant are crossed) is turned in with a throat-seizing. It is then secured to the clew-cringle of the sail (in bending) with a stout lashing, and the standing-parts of the sheets, after being hooked, or clinched, to an eye-bolt in each side of the bulwarks,† are rove from *out, in,* through a single block spliced into each end

*In merchant vessels, where the quarter-blocks are *threefold,* the top-gallant-sheet is rove in the *middle* sheeve.

†The position of the bolts and blocks (or sheeves) must be such, that the sheet, when taut, shall form a line at right angles with the luff of the sail—for otherwise, either the foot or the

of the pendant—the hauling-parts leading in-board on the forecastle to belaying-cleats.

NOTE.—Both parts of the topmast-staysail sheets are secured *inside* of those for the jib.

FLYING-JIB SHEETS.—These are single, and usually consist of one piece of rope, which is middled and fitted with a thimble, (like the pendants mentioned in the preceding paragraph,) and secured to the clew of the sail in the same manner. The parts are then led each side of the boom, in-board, through a thimble on the whiskers, (where whiskers are used,) and a fair-leader in the head rail, to the forecastle.

SPANKER SHEETS (Plate 31, Fig. 171).—These are invariably rove in *one* with the guy by either of the following methods, viz.: The standing-parts are put over the outer extremity of the boom with a cut-splice eye, or connected separately with sister-hooks, and rove to the (double) block in each quarter; thence through the sheet-blocks from *forward, aft,* and back to the second sheeves of the quarter-blocks. *Or,* the standing-parts, after being spliced around (and forming the straps of) the quarter-blocks, (which in this case should be single,) are rove from *aft, forward,* through the sheet-blocks; thence from *forward, aft,* through those in each quarter, to the guy-blocks, at the outer end of the boom, and then in-board again, to the second sheeves of the sheet-blocks.

BOOM-MAINSAIL SHEETS (Plate 31, Fig. 172).—In small craft, as schooners, &c., a single purchase of double-blocks, and working on a traveller, is used; but in larger vessels, two (attached by separate straps, and hooked to eye-bolts in each quarter) are employed to manage the boom—the hauling-parts in either case leading from the upper block. This latter method is by far the best, as every one who has had to "gibe" a boom-mainsail, with a single sheet and crotch-ropes in heavy weather, will bear witness to.

GAFF-TOPSAIL SHEETS (Plate 31, Fig. 174).—These are formed of a single piece of rope, which is middled, and the bight passed through the clew-cringle of the sail; the ends being thrust also through the bight, are led down on each side of the gaff to a belaying-cleat on the boom near the jaws.

STUDDING-SAIL SHEETS.—Those for the *lower* studding-sail consist of a single piece of rope, passed through the inner clew-cringle like those for the gaff-topsail; and in setting the sail, one sheet is rove from *forward, aft,* through a thimble or block on the goose-neck, in order to bring the clew close down to the boom, and the other led in-board on the forecastle to tauten the foot of the sail. (Plate 37, Fig. 189).

In fitting a *topmast* studding-sail, two sheets are also required, which are attached to the clew in the same manner as those for the lower studding-sail. One (called the *short sheet*) being passed *forward* of the topsail, and *aft* through a thimble (seized to the jack-stay on the outer quarter of the lower yard) into the top, where

leech, would become slack, and the jib thus be deprived of a great portion of its efficacy. Indeed, this consideration is too often overlooked in the equipment, much to the detriment of the sailmaker's reputation, and the "standing" of the jib.

it is belayed to a cleat; and the other, or *deck-sheet*, being led to the forecastle, *forward* of the yard.

NOTE.—In merchantmen, only a single sheet for the topmast studding-sail is used, leading to the opposite side of the forecastle—the sail being cut with a slope on the foot, which brings the clew below the lower yard.

The *top-gallant* studding-sail sheet is simply spliced into the clew of the sail, (having parcelling on it for two or three feet below, to avoid chafe from the foot-rope of the topsail-yard,) and led into the top, where it is hitched around the forward-swifter. In merchantmen, where the sail is more frequently set from the *deck*, the sheet is led there also. (Plate 37, Fig. 189).

TACKS.

FORE AND MAIN TACKS (Plate 30, Fig. 170).—These are invariably rove double, (except now and then in the merchant service, on board of small vessels, where they are single,) and fitted with the inner end tapering. The standing-part, which is wormed and served for a fathom or so from the end, (as a protection from wet,) is clinched, or passed over the extremity of the bumpkin* with a running-eye, and rove up through the tack-block at the clew of the sail—then back through a leading-block inside of the standing-part, and a hole in the bulwarks.

NOTE.—*Glascock* recommends (although not in very *intelligible* terms) a simple method of fitting the tacks for "singling" or "doubling" at pleasure, which consists in thrusting a toggle of *lignum-vitae*, into the standing-part, about two fathoms from the extremity. A piece of small-sized rope, of sufficient length, is then spliced into the reeving-end, to serve the purpose of a hauling-line, and when it is required to single the tack, for hauling up the sail, it is only necessary to cast off both parts. Again, in setting the sail, the 'hauling-part is run on board, the toggle taking the block, until a certain strain is brought upon the rope, when the standing-part of the tack, which is hanging downwards, immediately secured to the bumpkin, and the shell of the block relieved from the strain which the toggle would otherwise produce.'

'If it be required to set the sail, with a *double* tack, haul down on the small leading-line attached to the standing-part, and take a turn or two with it around the bumpkin. By this movement, the tack is doubled in a few seconds.'

STUDDING-SAIL TACKS.—These are either bent to the cringles of their respective sails with two half-hitches, or an inside-clinch, and rove from *in, out*, through the blocks at the boom-ends. That for the *topmast* studding-sail being led aft, through a tail-block on the forward-swifter of the main-rigging; and the tack of the *top-gallant* studding-sail, through a leader strapped around the lower dead-eye of the after topmast shroud; although in the merchant service it is oftener brought directly to the deck, and belayed well aft. (Plate 37, Fig. 189).

SPANKER AND BOOM-MAINSAIL TACK (LASHINGS) are passed through the cringle, (into which they are spliced,) and an eye-bolt on the upper side of the boom.

*The *main* tack is secured to a bolt and block in each of the water-ways, forward of the gangway, being rove like the fore, to the block on the clew of the sail.

NOTE.—In laying to, in a small vessel, under a balanced-reefed (boom)-mainsail, the tack of the sail should be lashed up to the jaws of the gaff, and the whole hoisted several feet up the mast by means of the throat halliards. In this way the sail is elevated to the wind above the waves, and in the event of being boarded by a quarter sea, it cannot lodge in the belly of the sail, but will pass between it and the boom.

JIB AND STAYSAIL TACK (LASHINGS) (Plate 33, Fig. 181).—These consist of a single piece of rope, spliced into the tack-cringle of the sail, and lashing it to the *stay*, around which it is passed, and the end hitched. This plan is preferable to that of lashing the tack to a hook or bolt *on* the *boom*, as recommended by Brady and others, particularly if we regard the jibs as *lifting* sails, and the head-booms as mere *outriggers*, which is manifestly the case. Besides, if the tack be lashed to the stay, much of the difficulty usually met with, in clearing the wreck of a broken jib-boom, will be removed.

CLEWLINES.

In this article it is thought proper to class all the running-gear partaking of the character of clewlines, as clew-garnets, clew-ropes, and clew-jiggers, giving to each a subordinate head, for ready reference.

CLEW-GARNETS (Plate 33, Fig. 183).—These are used only on the courses. The standing-parts are hitched to a thimble at each side of the slings of the lower-yard, for that purpose, and (best if) rove through "secret-blocks"* (Plate 33, Fig. 182) at the clews of the sail; then back to their respective leading-blocks underneath the yard, (*inside* of the standing-parts,) and lastly to the deck.

TOPSAIL CLEWLINES (Plate 33, Fig. 183).—These are best if fitted like the clew-garnets with secret-blocks, and rove through the forward sheeves of the quarter-blocks "via" the lubber's-hole to the deck. In many vessels, however, the method of a whip and pendant is preferred—the former (of *double* the sized rope for the ordinary clewline) being rove through the quarter-block, and attached to the clew-cringle with sister-hooks; while the parts of the whip (but *half* the size) reeving through a block, spliced into the end of the pendant, are led through the lubber's-hole to the deck, and separated widely from each other, on account of the liability to become "hawser-laid."

TOP-GALLANT AND ROYAL CLEWLINES (Plate 33, Fig. 183).—These are both single, and connected to the clews of the sails by sister-hooks, and rove through the quarter-blocks of their respective yards, and thence to the deck by way of the lubber's-hole; although in the naval service the royal clewlines do not extend below the top, being rove through fair-leaders on the topmast shrouds, alongside of the sheets.

LOWER STUDDING-SAIL CLEWLINES.—These are simply hitched to the clews, and reeving *abaft* the sail, through small single blocks on the inner end of each lower

*These are an admirable invention, consisting of a sheeve enveloped in a shell, and having holes in one extremity large enough to receive the rope—a plan which precludes the possibility of jambing the reef-points, in hauling up the clews.

studding-sail-yard, are thence led in-board, to a block hooked to the forward futtock-plate.

FORE AND MAIN CLEW-JIGGERS (Plate 32, Fig. 178).—These each consist of a purchase (as the term implies) of single blocks, hooked to the clews of the courses, and to eye-bolts underneath the forward part of the tops. In furling sails they are found very useful, for rousing the clews and leeches up forward of the yard; while they also serve the purpose of *inner* halliards for the lower studding-sails, and are often employed as yard-arm jiggers in bending, or reefing.

TOPSAIL CLEW-ROPES (Plate 32, Fig. 176).—These are an essential requisite in scudding, to clew up the sail, in the event of being caught aback, as also in furling. They usually consist of a pendant and whip—the former attached to the clew-cringle on the *forward* side of the sail, and rove through a block at each side of the topmast, under the eyes of the rigging; the whip reeving through a block spliced into the end of the pendant, and having both parts led to the deck, *via* the lubber's-hole.

NOTE.—In the frigate "Cumberland" the buntlines were double, and the *outer* legs (in clewing up the sail for furling, or what not,) were untoggled, and connected to the clew-cringles, in lieu of extra clew-ropes—a plan which answered every purpose.

BUNTLINES.

FORE AND MAIN BUNTLINES (Plate 32, Figs. 179, 180).—The neatest and best plan for fitting these (in a large ship—for example, a frigate) is as follows: Through a *double* block hooked at each side of the lower cap, and through two short leaden pipes, inserted between the tressle-trees and pigeon-hole, reeve both ends of each buntline, from *aft*, down forward of the sail to the foot-rope, where they are connected by toggles; having the bight of the middle part (to which a single whip leading through the lubber's-hole to the deck is attached) abaft the block at the cap. The other is rove in a similar manner. But in small vessels, where single buntlines are used, the cap-blocks are single also, and *one* leg only, is toggled to the foot-rope of the sail, some distance out on each side of the midship seam. The advantage of this plan is, that it enables us to rouse the buntlines up *higher;* while the objection to fitting the *main* by this method consists in the fact of crowding too many ropes together—a principle which should never be lost sight of in the reeving of running gear.

NOTE.—If it is deemed preferable to reeve the *main* buntlines, with the bight forward, through *shoe*-blocks between the stays, the tendency to become "cable-laid" can be obviated entirely, by having the *buntline* of *left*-hand rope, and the *whip* of *right*-hand.

TOPSAIL BUNTLINES (Plate 32, Fig. 177).—These are generally single, and the *fore* rove through the *lower* sheeves of fiddle-blocks* under the eyes of the topmast rigging—thence forward through the thimbles of a span, hitched around the breech of the tye-blocks, and down to the foot-rope of the sail, where they are connected by toggles—the hauling-part extending to the deck through the lubber's-hole.

*The *main* and *mizen* topsail buntlines reeve through cheek-sheeves in the fore part of the upper tressle-trees, or single blocks, at each side of the heel of the top-gallant-mast.

NOTE.—In many vessels the buntlines are fitted with whips, and are often spliced into short spans, which toggle to the foot of the sail, at each side of the midship seam.

TOP-GALLANT BUNTLINES (Plate 32, Fig. 175).—These consist of a single "leg," rove from *aft*, through a single block, under the eyes of the top-gallant rigging, (that for the *fore* on the *opposite* side of the flying-jib halliards,) and are spliced around the thimble of a span, toggled to the foot of the sail, about halfway in from the clew, at each side of the midship seam—the hauling-part leading to the top in men-of-war, and in merchant vessels to the deck.

NOTE.—A good plan is to connect it also with a span on the leech, about half-way down from the head-cringle, which (when the sail is clewed up) leaves it ready for furling.

BOWLINES.

(Plate 32, Fig. 180; Plate 34, Fig. 184)

FORE BOWLINES.—These are simply toggled to the bridles, on the foresail, and led in-board to the forecastle through blocks seized to the fore-stay straps.

MAIN BOWLINES—Consist of a whip and pendant—the latter reeving through the thimble in the bridle, and belayed to the fore bitts; and the former passing through a block in the end of the pendant led well forward—the standing-part of the whip being secured to an eye-bolt at the fore fife-rail, and the reeving end over a pin in the side rack. In tacking, when it is required to let go the main bowline, the standing-part of the pendant is cast off, and the whole shifted to the opposite side, ready for reeving.

TOP-BOWLINES.—The *fore* toggle to the bridles, and lead forward through blocks strapped to each side of the bowsprit cap, and back, in-board, to the forecastle. The *main* reeve through single blocks, connected to bolts in the after rim of the fore-top, and thence to the deck; and the *mizen* through the outer sheeves of the brace-blocks on the mainmast.

TOP-GALLANT BOWLINES.—The *fore* reeve through blocks seized to the jib-guys, and thence in-board to the forecastle The *main*, through the outside sheeves in the after-chock of the fore-topmast cross-trees; and the *mizen*, through the outer sheeves of their respective brace-blocks, to the deck.

ROYAL BOWLINES.—The *fore,* after being toggled to a *becket* on the leeches of the sail, are rove through thimbles on the royal stay about two or three fathoms from the *nip,* and thence through others on the guys, in-board to the forecastle. The *main* are rove through thimbles seized to the after-part of the fore-top-gallant funnel, and led from thence to the fore-top; while the *mizen* reeve through the outer sheeves of their brace-blocks, like the top-gallant bowlines, and reach to the top also.

NOTE.—In the merchant service, top-gallant and royal bowlines are invariably dispensed with as "extravagant superfluities."

DOWN-HAULS.

Topmast-staysail, Jib, and Flying-jib Down-hauls (Plate 35, Fig. 186).—These are each bent to the head-cringle of their respective sails, and after being rove through the hanks, and a single-block, seized to the nip of the stay, are then led in-board on the fore-castle.

Studding-sail Down-hauls.—That for the *topmast* studding-sail is bent to the outer end of the yard-arm, and rove from thence through a thimble on the leech to the down-haul block at the tack, leading on deck across the forecastle to the opposite side. That for the *top-gallant* studding-sail is merely bent to the inner yard-arm of the sail, and led abaft all to the top.

Gaff-topsail Down-haul (and Clewline) (Plate 35, Fig. 187).—This is led from the after-clew of the sail (to which it is bent) through a single block at the head of the sail, and thence through the hanks on the mast down to the deck.

OUT-HAULS.

Spanker Out-haul.—Vide Ibid. Booms.

Lower Studding-sail Out-haul.—This is clinched, or connected by sister-hooks, to the outer clew of the sail, and led through a single block, put over the boom with a rope-strap, either abaft, to a sheeve above that for the guy in the bulwarks, or in-board along the boom, to the forecastle.

Gaff-topsail Out-haul (Plate 35, Fig. 187).—This is hitched to the clew of the sail, and rove through a sheeve at the gaff-end, down to the deck, where it is belayed to a cleat on the boom.

Trysail Out-hauls (Plate 35, Fig. 185).—These are used only on board of those vessels where the trysails are set and taken in on the gaff. They are always single, and attached to the outer head-cringle of the sail, being rove through a sheeve in the gaff-end to a leader strapped to the jaws, and having a whip, which is led thence to the deck. The *in-haul* is rove through the hanks on the head of the sail to a single block at the jaws also, but on the opposite side.

REEF-TACKLES.

Topsail Reef-tackles (Plate 36, Fig. 188).—In the navy, they usually consist of a *hide* pendant, clinched to the cringle, on the leech of the sail below the close-reef-band, and lead upward through the *outer* yard-arm sheeve, to the lower block of the top-burton, which is hooked to a thimble spliced into the reeving of the pendant; the hoisting part of the burton-fall leading through the lubber's-hole to the deck. In the merchant service, however, where burton-pendants are unknown, and where the burton itself is often required for some other purpose, the reef-tackles are fitted thus: The standing-part of the pendant is spliced to the strap of a block toggled on the leech of the sail, below the close-reef-band—thence led upward through the forward sheeve of a double block on the pacific-iron—down through the block on the leech; up to the running sheeve of the double block, and so to the after sheeve of the quarter-block, and lastly, through the lubber's-hole to the *side* rack on deck.

Fore and Main Reef-pendants (Plate 36, Fig. 188) are simply clinched to the cringle, and rove through the outer yard-arm sheeve, or a block placed temporarily over the pacific-iron, being hauled up by means of the clew-jiggers, hooked to a thimble in the end, and a strap around the forward futtock-plate. In merchantmen they are often led through the after sheeve of the quarter-block, which is double, and roused up by a deck jigger attached to the end; and sometimes led like those for the topsails.

Boom-mainsail Reef-tackle.—Vide Ibid. Booms.

LEECH-LINES.

These are mostly confined to the courses, (Plate 36, Fig. 188,) and those on the *forward* side of the sail are clinched to the leech—the *outer* one about one-third down from the head-earing cringle, and the inner one about two-thirds—and thence rove up through leading blocks on the bending jack-stay* to the inner and outer sheeves (respectively) of a double block, strapped into the end of a long pendant, hooked at each side of the lower cap—the hauling-part of the leech-line reeving through fair leaders on the lower rigging to the *side* rack, on deck. The *after* ones are clinched to the sail in like manner, at the same points, but rove *abaft* through single blocks on the lower yard (corresponding with the position of those for the forward leech-lines,) to double blocks at the slings, and thence on deck to the topsail-sheet bitts.

Note.—In large ships they are sometimes temporarily rove on the topsail-yards (through tail-blocks on the forward swifters) for furling sails, where the leeches are heavy, and clew-ropes are not used.

BRAILS.

Spanker and Trysail Brails.—These are middled, and the bight secured to their respective eyelet-holes on the leech of the sail by cross-seizings—the *throat* brails reeving thorugh cheek-sheeves on the inner end of the gaff at each side, and the *foot* brails through single blocks seized to the hanks on the trysail-mast.

Jib Brails.—These are fitted in the same manner, and led in-board, except being rove through single blocks secured at each side on the luff, and through leaders between the guys and boom.

*These blocks should be so placed that the leech of the sails will be taut along the yard when hauled up; and fitted with straps, which permit them to hang about a foot below the yard—a plan obviating the necessity of attending the leech-lines in bracing up.

SLAB-LINES.

These are peculiar to the courses, and spliced into a thimble on a span toggled to the foot-rope at each side of the midship seam, and rove (abaft the sail) up through a single block, seized to the sling-band of the lower yard, or the span lashing of the quarter-blocks, and thence to the deck.

GUYS.

Lower-boom Guys.—When double, the standing-part of the *forward* one is put over the end of the whisker (when whiskers are used) *after* the jumper-brace, and rove through a single block on the boom, and back to another bolted to the bowsprit cap; the hauling-part being led in-board to the forecastle. When single, as in the merchant service, they connect to the boom by sister-hooks, and the block is dispensed with. The *after* guys are rove in the same manner, except abaft to a bolt and sheeve in each side of the bulwarks, just forward of the gangway.

Spanker-boom Guys.—Vide Ibid. Sheets.

BUNT-JIGGERS.

These are indispensable helpmates in handling the square sails, and are usually fitted thus: Those for the courses (though more properly bunt-*ropes* than jiggers) usually consist of a pendant, which is rove through a single block hooked to one of the upper links of the lower yard-slings, having sister-hooks in one end (by which it is connected to the glut on the sail,) and a whip on the other, leading to the deck. The *topsail* bunt-jiggers (of which there are two to each sail, a *long* and *short* one) consist of gun-tackle purchases, corresponding in length to these distinctions, hooked to a bolt in the topmast tressle-trees, at each side of the top-gallant-mast, and to the upper and lower gluts of the sail—the hauling-parts of both jiggers leading to the top. In the merchant service, however, where (as we have before had occasion to remark,) economy is the secret spring of all action and enterprise, in lieu of using two separate jiggers to rouse up the bunt, one is made to serve the purpose of both, by having a pendant spliced into the strap of the upper block, and reeving through a single one on the tressle-tree to the top. In the event, then, of requiring a long bunt-jigger—when, for example, the sail is wet and heavy—it is only necessary to ease off the pendant in the top, which permits the purchase to reach any distance. Again, when the short jigger is requisite, take to the slack of the pendant, and round up the jigger, holding the sail up on the yard, until the lower block can be hooked to the second glut.

As in all the *unwritten* sciences, the practical application of an important principle in seamanship, like that of which we have just been treating, is accompanied by an infinite number of *traditionary* precepts (so to speak) hoarded up in the mind of the practitioner. From these a few have been selected, which are offered as a concluding paragraph to this article—*not*, however, as infallible rules, attested by all the fluctuating circumstances of legitimate experiment, but rather as "helps and hints" in a professional dilemma.

BLOCKS.*—All those for *general* purposes should be in length *three* times the circumference of the rope rove in them; an exception to this, however, is met with in *clew*-blocks for the courses and topsails, which should be in length, but *once* the circumference of the required tack, or sheet.

Quarter-blocks, for the lower and topsail yards, &c., should be *twice* the circumference of the sheet that reeves through them.

Brace and *purchase* blocks, as they require a good sweep, ought to be in length *four* times the circumference of the rope. If attention is paid to this, the *blocks will give the size of the rope*, and conversely, the *rope the size of the blocks.*

BRACES, STUDDING-SAIL GEAR, &c.—Turn half the length of the *main*-yard into fathoms, and calculate as follows:

7 times half the length=the fore-brace.
6 " " " =the main brace=fore-topsail brace.
5 " " " =the main-topsail brace.
4 " " " =the cross-jack brace=mizen topsail brace.

Top-gallant braces (if double) should be of the same length as their respective topsail braces. If single, the same length, but *minus* the distance between the masts, (to which they lead,) and half the length of the top-gallant-yard.

The *size* of braces can be apportioned to very nearly, by allowing one inch for every 18 feet of the yard in length.

Lower studding-sail halliards=the mean length of the fore and fore-topsail braces.

Topmast studding-sail halliards=fore-topsail brace=topsail halliards, and sufficient over for grommet-straps to the blocks.

Topsail-tyes=length of their respective lower yards; and should be double the size of the halliards, for if the rope is doubled in size, it will give *three* times the weight, and consequently the strength.

SEIZINGS.—*Eye* do. for 11 inch rope, 21 thread; *Quarter* do. 15 thread.
" " 10 " " 18 " " " 12 "
" " 8½ to 8 " 15 " " " 9 "
" " 7 " " 12 " " " 6 "

*In the merchant service, blocks fitted with *friction* rollers have been *universally* introduced, and as a consequence it is a well-established fact that the number of men formerly required to man a vessel has been reduced nearly in the ratio of the decrease of friction. Yet, strange to say, this invention, tending so much to diminish the labor and requisite physical force, is scarcely known in the naval sevice, except from "hear-say evidence;" which, by-the-by, appears to be excluded with as much repugnance, as in any case of criminal law. And what's the result? That a merchant vessel, measured in *all* her equipment by the *universal* (and often *mean* and scanty) length of an individual *purse*, with a handful of men upon her decks, is frequently manoeuvred with as much facility as a man-of-war of equal tonnage, having *twenty times* the *personnel*, and replete with all the essential requisites of efficiacy and discipliine! And why? Because the spars and rigging are *necessarily*, in a great proportion, heavier, and the sheeve of *every* block on board, to use a vulgar phrase, "turns as hard as the bolts of Newgate."

GUN GEAR.—In the absence of a better, the following "rule of thumb" will be found to answer for all practical purposes: *Breechings* should be *three* times the length of the gun, if fitted with shackles; but where they clinch or knot outside, (an old and *senseless* plan,) the thickness of the bulwarks must be added thereto.

In calculating the quantity of rope required for the *side and train tackles* of a battery, allow *five* times the length of the gun for each purchase of double and single blocks; and for every one with two single blocks, *four* times that quantity. This will afford sufficient for grommet-straps to the blocks But where they are already fitted, allow *four* times the length of the gun for the first case, and *three* times for the second.

FITTING AND BENDING SAILS.

Prior to the performance of this important feature in the equipment, the sails should be carefully examined, in order to supply any omissions of the sailmaker; and it may not be amiss to name the points, clews, cringles, bridles, thimbles, beckets, eyelets, toggles, and gluts, as parts necessary to include in this precautionary scrutiny. In addition to which, the fore and aft sails must be prepared with hanks, brail-blocks, lacings and lashings, and the square sails with earings and rope-bands. The *earings* (divided into two classes of reef and head-earings) should be spliced into eyelets, just below their respective cringles, with an eye sufficiently long to take one outer and one inner turn, which are consequently equal to two *single* outer and inner turns. The reeving ends of the *reef*-earings (which are whipped) being made fast to the cringles of the next band (*above*) with a bowline hitch: for example, that for the close-reef to the third; the third to the second, and so on; the object of which is to place each earing within the reach of the men in reefing; a simple measure, but nevertheless of some magnitude in practice. The *rope-bands*, consisting of two-yarn foxes, are middled, and secured to the head-rope, by thrusting *one* end through the bight, which is first passed through the eyelet from the *fore* side of the sail, and hauled taut. The next consideration is to place the *gaskets* on the yards. These are classed as bunt, yard-arm, and sea gaskets, and are usually made of plaited yarns. Those for the *bunt*, consist (in the navy) of two single legs—one on each side of the slings, varying from two to three inches in width, and fitted with a thimble in one end, by which it is secured to the *bending* jack-stay with a permanent seizing—the other extremity having a laniard, which is hitched to the opposite quarter of the yard on top; the gaskets crossing each other on the bunt when the sail is furled. This plan, however, is not so common in the merchant service. There the midship gasket more generally consists of a skeleton triangle of sennet, with a thimble in each corner—those at the base being secured to the jack-stay by seizings, and the upper one (which is fitted with a laniard,) hitched to the tye or yard-slings, as the case may be. The *yard-arm* gaskets are made of sennet also, and fitted with a thimble, or eye, in one end, and the other tapering, and secured at equal distances (generally about every third seam) along the yard, *underneath* the jack-stay, by a cross-seizing just below the thimble. The *sea-gaskets*, or more properly furling lines, (of which there are two on each arm of the lower and topsail yards,) may be either of sennet or small-sized rope, and of sufficient length to go around booms and all, when furling in heavy weather. These, however, are not necessarily permanent fixtures to the yard, although usually put round it at the outer and inner quarters with a running-eye, and the surplus end bighted up with frapping turns, and thrown forward of the sail, at sea.

In order to illustrate more fully the minutiæ of bending sails, it is thought proper to bend each separately, and in the order of its size and importance, without regard to the usual routine, even at the risk of in*cur*ring censure from the *critiques* of the "manger" for that repetition of terms and technicalities, so characteristic of professional books.

COURSES.

Stretch the head of the sail across the deck forward of the mast, having the *roping* on the *under side*, and gather the leeches up in bights, by bringing the reef-bands successively to the head-rope, which will leave all parts clear for bending on the rigging. Toggle the buntlines to the foot-rope at each side of the midship-seam, and clinch the leech-lines to the cringles, stopping both to the head of the sail—the former to the eyelet of the middle rope-band, and the latter in the wake of their leading-blocks. Hook the clew-jiggers* from straps around the pacific-iron to the first reef-cringles, hitching the head-earing to the standing-part. Shackle or *seize*† on the clew-garnet blocks (with the ropes rove) to the after part, passing them *under* the sail. Send hands on the lower yards—trice up the booms—man the gear and sway aloft, merely gathering up the slack of the clew garnets. When high enough, (i. e. when the centre of the sails reaches the centre of the yard,) bring to, cut the buntline and leech-line stops, and make fast the midship rope-band, together with three or four others at each side of it, by passing the short ends *under* the jack-stay from *forward, aft,* and the long ones *over* and *under*, from *aft,* forward—thence back through the eyelets, and square-knot them on top. Haul the sail out until the head-rope is taut along the yard, and pass two or three turns of the earing through the thimble of its strap, and if the sail is new, let the men ride it down in order to stretch the head-rope, and bring the cringle and thimble as nearly together as possible. Then pass the earing afresh, and for a full due—viz. *two* turns from out, in, through the thimble and back to the cringle, expending the end leading last from the latter, *over* and *under* the yard, *inside* of the head-earing strap, and secure it by hitches around the turns *first* taken. Now make fast the remaining rope-bands in the same manner as the midship ones, and clinch or reeve

*Where the reef-pendants (or properly tackles) are rove, as in the merchant service, viz. from the yard, through a block toggled to a short bridle on the leech, below the last reef-band, the clew-jiggers may be dispensed with, and the sail hauled out to the yard-arms by these pendants, which in this case should be rove on deck, and the head-earing cringle stopped to the standing-part. This plan (which was adopted on board of the "Columbus," 74,) apart from the facility it affords in reefing, will be found serviceable in bending and unbending the sail, whenever it may be essential to perform that evolution.

†In lieu of having the shackle large enough to receive the clew-garnet block, with those for the tack and sheet, as recommended in the note on page 56, the former is sometimes *seized* to the clew as here stated, for the reason, that if the shackle should part or unkey, the sail may still be hauled up by the clew-garnet.

the reef-pendants, as the case may be. Haul up the clew-garnets snugly, and shift the yard-arm jiggers to the forward futtock-plate, for the clew-blocks, (through which the tacks and sheets are rove,) and shackle them to the cringles. Lastly, slack down and overhaul the gear, and set the sail to ascertain its fit, and if all is rove right—then clew up—hook the bunt-jigger, pass in the leeches, and furl.

TOPSAILS.

Having prepared the sails for bending, and supplied any deficiencies which may be requisite, place them on deck forward of their respective masts, rolled up on the *fore* part,* and secured by three or four rope-yarn stops on each side, having the first reef and head cringles together, and the foot-rope clear for bending on the rigging. Hook the sail-tackle† from a strap on the topmast-stay collar, to *one* bight of a stout pair of slings, passed around the centre of the sail, and *seize* the other to the *back* of the *hook*. Overhaul the sheets and the reef-pendants, bringing the bending-ends of both into the top, and have the bowlines, buntlines, and clewlines ready for toggling, and bending to the sail. Send hands aloft on the yards; man the tackle—trice up the booms, and sway away. When the sail rises above the top, cut the stops; shackle or clinch the sheets and reef-pendants, stopping the head-cringle to the latter, into the thimbles of which, hook the top-burtons.‡ Hook also the clewlines (or if double, *lash* on their blocks) to the clew-beckets, and toggle the buntlines and bowlines to the foot-rope and bridles respectively. Now man the burtons, and sway up on the tackle, until the centre of the sail reaches the centre of the yard, then pass the middle rope-bands; rouse out the head-cringles, cutting the seizing of the reef-pendants, and that of the slings around the sail; at the same time gathering up the slack of the buntlines and clewlines. If requisite, ride down the head-rope, and pass the earings, (for a full due,) and the remaining rope-bands, as in the case previously stated. Then unhook the burtons and the tackle from the stay—overhaul the gear; sheet home the clews; down booms, and hoist the sail up to a taut leech, both with and without the reefs; after which, lower the halliards; clew up—hook the bunt-jiggers, and furl.

JIB.

This sail, as with those previously mentioned, requires certain equipments, not strictly within the contract of the sailmaker; and as such, mingled with the varied and comprehensive duties of the seaman. Among these "extrinsic requisites," we may note the lashing on of the *bonnet, hanks*, and *brails*. The former, although confined chiefly to the naval service, and to small craft, as pilot-boats and sloops, (in which a staysail is not used,) is secured to the jib as follows, by a lacing spliced

*The reason for making the sail up on the fore part is to cause it to fall *forward* and clear of the top when cast loose from the slings.

†This consists of a long stout burton, for the purpose of bending sails, and is allowed to all the vessels of the naval service.

‡On board of merchantmen, in lieu of the burtons, the sail is hauled out by the reef-tackles alone, the head-cringles being stopped to the standing-part, as in the case of the courses.

into the tack-cringle of the sail: Pass a *bight* of the lacing (from *starboard* to *port,* for example) through the forward eyelet of the bonnet, and up through the corresponding one of the jib, from *port* to *starboard;* then reeve the *end* through the bight, and haul it taut; after which, form another bight as before, and proceed in the same manner, reeving it alternately in the eyelet of the bonnet, and that of the jib, until all are included, and secure the end by two half-hitches to the cringle of the leech.

The *hanks,* which usually consist of stout thimbles sufficiently large to traverse freely up and down the stay, are attached to the luff by good-sized foxes of spun-yarn, rove through the eyelets, and passed around the hanks on *both* ends, which (being of sufficient length to fill the score) are knotted on the after side, and secured in frapping turns, between the luff and hanks.

The *brails* (though previously described in the article on "Running-rigging") are simply middled and secured to the leech eyelet with a cross-seizing; after which, they are rove forward through single blocks, fitted with a short-span, and seized to an eyelet, or cringle on the luff, for that purpose. In merchantmen, however, brails are seldom used on the jib; and many seamen, who deem them "superfluous gear," assert that the chafe induced by brailing up is sufficient of itself to exclude them from the calendar of "necessities," and to even justify the hauling down and hoisting of the sail, whenever it may be desirable to dispense with it, or essential to shift the sheet. (*Quad erat demonstrandum.*)

Now in order to bend the jib, cast off and unreeve the stay from the boom, (having a marrying-line bent to the becket,) and bring it in on the forecastle, or either side of the topmast-stays. Make the sail up in *bights* on the *foot,* in such a manner that the hanks shall stand one *above* the other, and reeve the stay *down* through them, and the down-haul *up,* hitching the latter to the head-cringle of the sail, and seizing both to the lower hank, just above the *nip.* Reeve the brails, and pass them (after the manner of a lashing) around the sail, and hook the halliards into the bights, swaying it up above the bulwarks. Re-bend the marrying-line, (the hauling-end of which is also in-board,) and man it, together with the down-haul, rousing out the jib, until the nip comes to the sheeve in the boom. Then clap on the luff-tackles, and set up the stay, for a full due; cast off the brails, and unhook and *re*-hook the halliards to the head-cringle—reeving the former through their leading-blocks (between the guys) in-board to the forecastle. Seize on the sheet-pendants to the clew; man the halliards; cut the racking of the lower hank and down-haul, (passing the latter through its block, which must be seized to the nip,) and hoist the sail up to a taut luff. Then lash the tack to the stay; trim aft the sheets; work the brails, to ascertain if all is rove right, and haul down and furl the sail.

TOPMAST-STAYSAIL.

This sail is made up like the jib, viz, with the eyelets and the head, tack, and clew cringles clear for bending; and is swayed out on the bowsprit, by means of the down-haul and halliards—the latter being hooked to a stout strap, passed around the body of the sail, in lieu of brails, (of which there are none.) The hanks, which are either small wooden hoops, or horse-shoe thimbles, having small eyes at each side of the opening, are secured to the eyelets by means of rope-yarn foxes. In the merchant service, where the topmast-stays are fitted to go around the bowsprit, as alluded to in the article on "Standing-rigging," the two are seized together at the tack of the staysail, and the hanks made large enough to traverse on both.

FLYING-JIB.

The routine of bending, as stated in the case of the jib, answers for the flying-jib; there being a few peculiarities only, to which it is perhaps well to refer, viz., it has neither brails nor bonnet; and in bending, the sail should be sent out on the *starboard* side of the bowsprit, on account of the inclination of the wythe (and consequently the boom) to that side.

NOTE.—Much of the labor incident to the bending of the sail, may be obviated by fitting it with a lacing in lieu of hanks,—a plan which is infinitely preferable for neatness, and which saves the necessity of turning out the stay. This lacing is hitched to the upper eyelet, and rove in every alternate one to the tack-cringle, where it is secured—leaving between the eyelets a surplus bight, which (when it is required to bend the sail) is passed around the stay, *against the lay*, and seized to the eyelet omitted in reeving.

While writing of this sail, it is thought proper to call the attention of the reader to a late method of fitting the sheets with two single blocks (through which they are rove) at the clew, and the standing-parts made fast to the lower end of the dolphin-striker. This plan, it will be perceived, corresponds, in some measure, with the principle laid down in the note on page 57, in reference to the position of the bolts.

ROYALS AND TOP-GALLANT SAILS.

In reference to these sails, little need be said, (except that they should be always bent on deck, on account of the difficulty of hauling out by hand,) as the earings and rope-bands are passed like those for the courses and topsails; the buntlines, bowlines, clewlines, and sheets, being bent after the yard is crossed. If, however, it should be necessary to bend the *top-gallant*-sail aloft, as is frequently the case on board of merchant ships, it may be sent up by the yard-rope, and the head-cringles hauled out by means of the top-gallant studding-sail halliards.

NOTE.—In furling either a royal or top-gallant sail, it should be made up with a *long, taut bunt*, and the clews "tucked in," to avoid tearing the sail in its upward or downward passage.

FORE AND AFT SAILS.

The manner of bending these sails is so nearly identical, that to save room and avoid repetition, it is deemed proper to class the spanker and trysails under this head, and to bend them in the old-fashioned way—referring in the end to the different methods of equipment, and to the peculiarities of each.

In the making up, there appears to be no particular rule of guidance, except that each sail must be placed fore and aft, under its respective gaff, (which should be lowered down for the purpose,) having the head and luff clear for bending. Reeve the brails in their blocks and sheeves, and pass the sail through the bights, rousing up and securing the *knock* (or throat) earing, which is usually passed around the gaff, and through the cringle. Although a better plan, and one frequently adopted in the merchant service, is, through a score (cut fore and aft in the gaff, directly between the jaws) and an eye-bolt on the *upper* side—the end being secured by hitches around the turns, of which there are generally *three*. Now haul the head of the sail taut along the gaff, using a jigger for the purpose, if requisite, and pass the peak-earing, (consisting of small-sized ratlin stuff well stretched,) which is taken (in *two* turns) from the cringle over the gaff, *abaft* a cleat on the upper side, and back to the cringle; expending the remaining part through the cringle. and around the gaff *inside* of the cleat; after which, take two half hitches around the outside turns underneath, to bind all parts together, and stop the end. Seize on the brails to their eyelets on the leech, (or, if the sailmaker should have omitted to work them, the point for those of the throat and peak may be obtained by bringing the leech in taut along the gaff, and marking the contact of the blocks and sheeves;) then reeve them, and pass the head-lacing from *aft, forward,* through the eyelets, and around the gaff, (taking a marlin-hitch at each,) on the upper side of the latter. Lastly, man the halliards, keeping the peak a little in advance of the throat, (in order to hoist easier,) and seize the luff to the hoops (or hanks) on the trysail-mast,* as the sail goes up, by passing the rope-bands (which are middled in the eyelets) on both ends, and knot them in frapping-turns, between the luff and hanks. As a *finale,* (if the sail is a spanker,) clinch, or hook the out-haul to the clew-cringle, and reeve it in its sheeve. The tack is then either lashed to a bolt on the boom, close to the jaws, or fitted with a small gun-tackle purchase, hooked to the deck, and brought down taut along the trysail-mast.

NOTE.—In vessels where the spanker and trysails are fitted to traverse on a stationary gaff, as referred to on page 47, they are usually sent up for bending by means of the in-and-out haulers, stopped to the throat and peak-earing cringles, and rove in their respective leading sheeves. Brails are then dispensed with, except one for the foot; and the head of the sail is secured to the hanks on the gaff, as in the case previously stated—the luff being seized to the batten or jack-stay, with spun-yarn foxes.

BOOM-MAINSAIL.

The routine, as stated in the preceding article, will be found to answer for this sail, there being, however, necessarily some slight exceptions, viz., the omission of brails; and in lieu of an out-haul, the clew is shackled to a span on the boom. Another distinction is, that the last three or four hanks are omitted, and their absence supplied with a lacing, passed *before* the mast, and through the eyelets instead of a round turn—which latter plan, though frequently practised, is nevertheless ill-suited to tricing up the tack, on account of binding.

NOTE.—A small single block should be seized to the middle hank on the mast, for the tack tricing-line.

*To the *travellers,* if fitted railway fashion.

STUDDING-SAILS.

The following method of bending these sails (though requiring more time than is usually bestowed) will be found the neatest and best, and in the end to fulfill all the conditions sought for, by extra pains and care—the sum and substance of which, may be reduced to the simple consideration of having the roping of the sail on the *after* side of the yard in lieu of underneath, and secured in such a manner as to preclude the possibility of its "bagging down."

No. 24

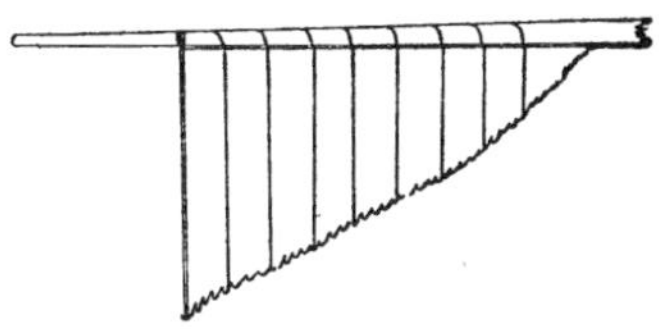

The outer earings, which are spliced into the cringles with a short eye, are passed through holes bored in the extremities of the yard, from the after side—thence back through the cringle and over the yard, *inside* of the hole, until three or four turns are taken, when the end is hitched through the cringle, and around the single part, as represented in the accompanying figure. The sail is then brought taut along the yard—the inner earing passed in the same manner, and the head-rope secured by a neat lacing of housline, rove around the yard and through the eyelets with a marline-hitch on top, at each of the latter. Lastly, the gear, or at least that portion of it which includes the sheets and down-haul, is bent after the manner described in "Running-rigging," and the sail made up, on the after side, by bighting the leeches towards the head, and running the gear along the yard.

In conclusion, many may find fault with the absence of a regular routine of bending the sails together, but this has been omitted as superfluous, in order to state clearly every point in reference to the subject; because it is a knowledge of the *minor features* alone, in any profession, which constitutes the *sum totum* of successful practice. And it seems but reasonable to maintain, that the mind, which is thoroughly cognizant of *every* peculiarity in the minutiæ of a branch like this, (where there is a mutual and characteristic dependence,) can never be at a loss to adapt its faculties to a combination of that knowledge in *one* system of practical application.

GROUND TACKLING.

Under this general technical term is comprehended all those multifarious appurtenances to the anchors and cables of a ship, known as stoppers, nippers, messengers, hook-ropes, clear-hawse pendants, deck-tackles, cat-falls, fish-davits, buoys, chain-hooks, compressors, &c. And for the sake of greater perspicuity, it will be necessary to class each one under a separate subordinate head, in the order of its importance; and to accompany each classification with a statement of its use, and method of equipment, both in the navy and mercantile marine.

CABLES.

The advantage resulting from the introduction of iron, as a substitute for rope, in the equipment of vessels, is nowhere so forcibly illustrated as in the use of *chain*-cables; and perhaps the best evidence which can be adduced of their superiority over the hempen ones of the "dark-ages," is the universality of their adoption. Under these circumstances, little need be said of the latter, except so far as it may be necessary to refer to some important feature, resulting from that connection always existing between an original and its successor.

The first consideration is, that the chains should be clearly and distinctly marked—for although having a shackle at every fifteen fathoms, and a swivel at the forty-fifth, or sixtieth, (depending upon the length,) it is essential to have not only these distinguished by some mark, but other points of the cable also, in order to know at all times, while at anchor, the amount by which the ship is riding. In the naval service various methods have been adopted, but as yet no *perfect* system arrived at, beyond placing small wire rings around the *stretchers*. *Totten*, who is perhaps the best criterion, thus states the routine which is pursued in the United States Navy: "After the first shackle, which will be fifteen fathoms from the ring or the anchor, measure off five fathoms, and mark with two turns of wire for the twenty-fathom mark, one for the twenty-five, three for the thirty, and so on."

In the merchant service, however, even this system, replete as it is with imperfections, is not practised; and naturally enough, where the *chronometer* is sometimes set by a foundry-bell, or an evening-gun,* the amount of chain in the hawse is invariably calculated by that *anti*-mathematical system of *guessing* so available to Yankees in a "tight-place."

Having marked the chains, proceed to get them on board. The manner of doing this, however, will depend in a great measure on the position of the ship. If moored in the stream, as we have thus far supposed, bring them alongside in a lighter or scow, (ranged fore and aft, in order to divide the pressure equally upon the deck or flooring,) and take them in through the vacant hawse-holes, with the assistance

*This assertion, though somewhat comprehensive in its bearing, is by no means applicable to the whole fraternity of merchant sailors; for the high character and perfection of our commercial marine are an evident contradiction to it, and sufficiently palpable to every one, to confine a statement of this nature within *appropriate* limits. The following dialogue, however, which took place about twilight (*not* a hundred years ago) between the officer of the deck, on board of the frigate C———n, at anchor in Callao Bay, and a *whaling* captain, whose vessel was weighing under the stern of the frigate, will not only serve to illustrate the truth of the remark, but to point out its application:

"Ship ahoy!" bawled the captain.

"Halloa! what do you wish, sir?"

"My *ker*nometer's down, and I want to know, stranger, what time you fire that gun of you'rn?"

"Eight o'clock, sir," responded the lieutenant, and down popped the "capting" on his "*winding* way," to the delinquent timekeeper of "Frodsham & Co."

of deck-tackles and chain-hooks. If, on the contrary, the vessel should be secured to the wharf, they may be got in either through a plank *shute*, leading from one of the midship ports to the chain-pipe, (in the locker;) or over a half-rounded spar, lashed on top of the gunwale. Prior to the stowage of the chains, however, it becomes necessry to secure the end below, as a preventive from loss, in the event of being unable to check its outward passage in veering; and perhaps the best method for accomplishing this object is the following, which was adopted on board of the frigate "Cumberland:" Through a ring-bolt in the kelson, the end of the chain was rove up to an iron roller, attached to the beams of the lower deck, immediately above—the last link of the chain being curved, in order to fit over a short perpendicular arm on the surface of the roller, which was kept from turning by a check-lever, having a small tackle attached, as reprsesented in the subjoined figure. In the event, then, of having to *slip*, it only became necessary to haul on the jigger, which permitted a revolution of the roller, and disengaged the link from the arm.

No 25

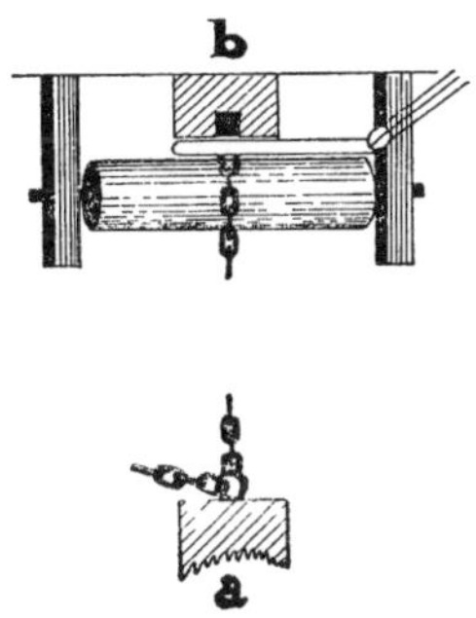

The simplicity and security of this plan cannot fail to strike every intelligent reader. In the merchant service, the usual method of securing the cable is by means of a lashing either through a ring-bolt in the kelson, and the last link, or around the heel of the foremast.*

The stowage of the chains is effected by ranging them *fore and aft*, assisting to do so, if requisite, with a hook-rope, (which is used to form the bights,) and rove through a tail-block in the after-part of the locker.

BENDING HEMP-CABLES.

Reeve the starboard ring-rope (from the block on the anchor-stock) through the ring from *in, out*, having passed the end in through the sheet hawse-hole, and bend it on the cable with a rolling-hitch, four or five fathoms from the extremity—putting on stops as requisite. Pass the standing-part of the fore-bowline in, and bend it to the cable also, in the same manner. Then with the latter rouse it out, having *slip*-ropes at the fore-tack bumpkin and cat-head, to keep the cable clear of the bow; as also a hand on the bowsprit to cut the stops of the bowline clear, as the cable comes out to the block. Haul the end aft, and through the ring of the

*For many obvious reasons the chain-lockers of merchant vessels are placed well forward, and as near the windlass as possible.

anchor, by means of the ring and slip ropes, (casting off the bowline,) until a portion of the cable equal to the length of the shank is roused through the ring, at which it should be seized—then form the clinch "*with* the sun," having the end *inside*,* and clap on the bends, which consist of ratlin-stuff middled and passed on the bight around both parts of the cable—cutting the stops of the ring-rope, and rendering the cable through the clinch, in order to make it as small as possible to avoid chafe from the shank-square.

The *port* ring-rope is rove in the reverse manner—viz., from *out, in*.†

The following plan for hemp-cables, recommended by Lieut. Fordyce, and practised in the British Navy, will be found to save much of the time and labor incident to the operation of bendng, as detailed in the foregoing paragraph: Unlay the strands of the bending-end of the cable about three fathoms, and make a long splice with two of them—keeping in the lay. Then with the remaining strand make an eye-splice to come fair with the bight of the two strands, and get the cable on a stretch—opening well the eyes thus formed with a fid. Tail stuff on to the ends of the eye-splice, and worm it four or five fathoms down the cable, clapping on stops every four feet or so to the end of the worming. Remove the fid, and thrust in an oval cut-thimble large enough to receive the pin of a chain-shackle, having it well tarred and parcelled, and lay a strand of rope in the cut-line, where the two eyes come together, and put on a stout seizing close up to the thimble over strand

No. 26

and all. This is called an "Elliott's-eye," and when the sheet-cable is required for bending, it is simply shackled to a chain length, attached to the anchor in the usual way, either as a permanent fixture or otherwise.

NOTE.—In shackling chains together, it would be well to remember that the *round* part of the shackle should always be placed *forward*, so as to pass out of the hawse first: by this precaution is can never get foul in running.

ANCHORS.

The method of getting on board the anchors in merchant vessels, is so nearly similar to that adopted in men-of-war, that it seems scarcely necessary to draw a distinctive line between the practice of both, although the absence of "sheets" in the former, and the necessity of heavier anchors in the latter, (arising from the peculiar design and service of war vessels,) produce a slight difference in the purchases used.

CAT-FALLS.—These consist of a double, treble, or four-fold purchase, (depending upon the relative size of the vessel and anchor,) rove and fitted as follows: The standing-part of the fall, after being secured to the cat-head either with a running-eye close up to the rail, or by means of a stopper-knot through a hole just forward

*This is technically termed a jambing clinch—and as it is designed *not* to *slip*, the reason is obvious.

†Much of the misapplication incident to similar expressions, as "in, out," and "out, in," may be obviated, (in this particular case at least,) by remembering that *both* ring-ropes are rove *from port* to *starboard*.

of the sheeves, is passed down through the forward sheeve of the cat-block, (which is iron-bound, and fitted with a hook, large enough to receive the ring of the anchor;) thence up through the corresponding sheeve of the cat-head, and so on alternately to the after sheeve of the latter, from whence it is led in-board through a leaden pin in the bulwarks, or simply over the rail. In many vessels this is rove in the reverse manner, viz. from aft, forward.

FISH-DAVITS.—In all the vessels of the naval service (above sloops-of-war in size) a perpendicular davit of oak, temporarily stepped in a saucer immediately abaft the bill-board, is employed to fish the anchor, using it on either side. On board of the "Cumberland" this was rigged as follows;* the davit being made with shoulders on the upper end for the gear, and the heel rounded to correspond with the concave mortise of the saucer—in which it was stepped by means of the pendant-tackle, hooked to a strap around the head.

No. 27

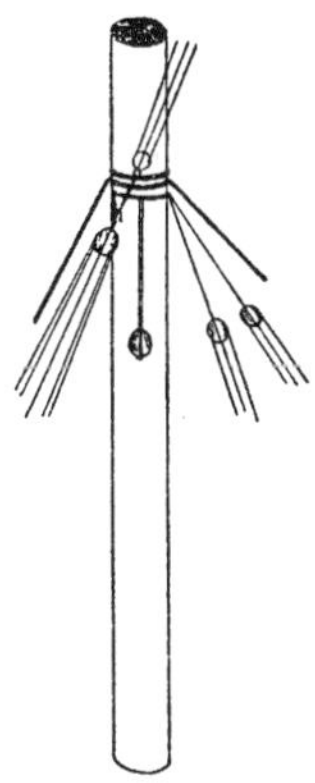

1. *Fair-leader.*—This consisted of a stout single block, (for the hauling-part of the fish-purchase,) fitted with a long double strap, which was served and placed over the davit-head first of all.

2. *Guys.*—These, one forward and one aft, were single pieces of heavy rope, going over with round spliced eyes—the former securing to a cleat on the cat-tail, and the latter setting up to a bolt in the fore channels.

3. *Fish-fall.*—This (consisting of a double and single block—the latter fitted with a large iron-hook like that for the cat) was attached to a stout strap around the davit-head, as illustrated in the foregoing figure—the hauling-part of the purchase leading up through the fair-leader, and connected to the lower block of the pendant-tackle, by means of a backwall-hitch.

4. *Topping-lift.*—A small stout jigger, (used to cant the davit in-board or out-board,) hooked to a thimble on the upper extremity, and a strap around the fore-mast, some distance above the deck.

5. *Bill-tackle.*—This, consisting of a similar jigger fitted with a long pendant and eye going over the davit-head, was hooked to one of the lashing-bolts in the bill-port, and used for canting the fish-purchase.

*The rigging of this might be greatly simplified by having a shoulder-band with eye-bolts for the guys and topping-lift.

6. *After-tackle.*—This consisted of a jigger also, connected to the davit-head, as in the previous case, and to a bolt in the deck, at the partners of the foremast.

In the smaller class of war vessels, in lieu of a perpendicular davit, a stout spar, (with a sheeve in each end,) placed horizontally athwart the deck, and projecting through the bill-port, is used, having an in-and-out jigger attached to the heel, and the fish-*pendant* (fitted with a stout hook in the lower extremity) led through the outer sheeve of the davit, up to the lower block of the pendant-tackle, by means of which the anchor is fished. On board of merchant ships, a plan differing from both of these is pursued, which saves the necessity of davits, by having a long stout purchase at the fore-topmast-head, hooked into the end of the fish-pendant, which is passed down over the rail, outside of all.

BACK-ROPES.—These are simply small leading-lines, either grafted or hitched to the back of the cat and fish hooks, and of length sufficient to reach from the rail to the water-line, with ease.

CAT-HEAD AND SHANK-STOPPERS.—The following diagram will serve to illustrate a recent invention, by which both the ring-stopper and shank-painter are cast off simultaneously. The former, which is of chain, secured by a stout shackle to the forward side of the cat-head, is passed through the ring of the anchor, (when catted,) and the last link placed over a hinged-shoulder, maintained in an upright position by means of a hook-lever extending across the cat-head. The shank-painter is secured in a similar manner at the bill-port. To each of these a trigger is attached, fitted with a small chain leading in the water-way to the arms of a swivel, worked by means of a bar shipped in the mortise (*c*)—giving it a half turn, either to the right or left, which disengages both stoppers at the same instant.

No. 28

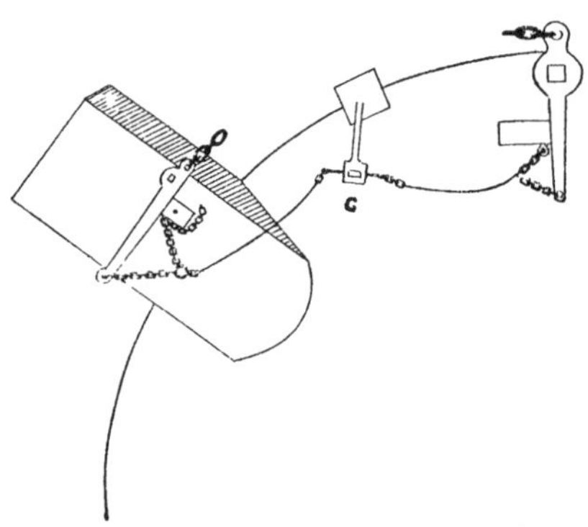

The worth of this invention will be apparent when we consider that in addition to its advantages in letting go the anchor, it admits also of being unshipped, and stowed away when not required for use—in which case the stoppers are belayed around cleats, or timber-heads, and the anchors lashed. The plan hitherto has been, to have the shank and ring stoppers of rope and chain, simply taken over cleats or cavils, and let go by hand, at the conclusion of the words "one—two—*three!*" Subsequently this was modified by the introduction of the triggers, which were (and are now in many vessels) set off by means of a top-mall, and a hand at each. Later again, however, this was greatly improved by Commodore Jos. Smith of the United States Navy, to whom the addition of the swivel and leading chains is due.*

Having rigged and rove the cat and fish falls, overhaul them down, and prepare for getting up the *bower* anchors, which should be brought under the cat-head in scows, with the crowns aft.† If the chains are already stowed, as we have supposed, rouse up the ends, and pass them out of their respective hawse-holes, (the forward or inside ones,) shackling them at once to the anchors. Hook the cat—man the fall—dropping the scow out of the way, and walk the anchor up—passing the ring-stopper, as the fall comes two blocks.‡ Now rouse in the chains until hand-taut, and mark them in the wake of the hawse-pipe for cutting, which "will save much of the customary noise incidental to the cry of cable enough."§ Hook the fish (from *out,* in) to the shank—man the purchase and sway away, until the palm of the upper fluke rests against the rail, or bill-port—then pass the shank-painter, as in the case already stated. The davit having thus performed its office, is unstepped by means of the pendant-tackle—the rigging taken off, and all stowed away.

Finally, bitt the cables by throwing the bight (formed of the slack chain abaft the bitts) over the bitt-head, so that the after part shall lead from the cavil to the locker, *outside* of the other part. In merchant vessels this operation is necessarily quite different, on account of having a windlass; the chain being passed over and around the *welts,* with one and a half, or three turns, as the case may be, prior to bending to the anchor.

Note.—In the new steamers "Atlantic" and "Pacific" the hawse-holes are cut on the main deck, and the chains led by means of iron pipes up to the spar deck, on which they are worked.

We come now to a consideration of sheet anchors—a subject scarcely concerning the merchant service at all, since they do not fall within the pale of "allowances." In the cases of ordinary practice, the following routine is adopted in the navy: Having the anchors alongside in lighters or scows, with their crowns aft, and the shores and stoppers in readiness, prepare to secure the yards, and reeve the purchases. Supposing it to be the starboard waist anchor, brace the main-yard *up* and the fore-yard *in,* in such a manner as to render the action of the purchases at the stock and crown as nearly vertical as possible. Then top up on the lifts on the side used to give the yards an inclination of about 10° with the horizon. Overhaul and hook both top burtons from the lower and topmast heads respectively, to straps around the yards on the starboard side—one within and one without the anchor-purchase—hauling them taut on deck, as also both lifts and the trusses—observing the precaution that the starboard lifts and burtons bear an equal strain, and that

*This improvement was first put in practice on board of the "Cumberland," at that time bearing the flag of the Commodore, while in command of the Mediterranean Squadron.

†Where anchors are brought alongside in this manner, to be got on board, it must be remembered to stow the *starboard* one on the *port* side of the scow, and *vice versa.*

‡The stopper is first hauled taut through the ring of the anchor, by means of a jigger, and three turns, one above the other, taken around the timber-head or cleat, and the end stopped to the first turn with a temporary seizing.

§Glascock's "Manual."

the *opposite* braces are not taut. Lash a large bull's-eye on the *after* side of the yards far enough out to permit the purchases to hang clear of the ship's side, and plumb over the anchor berth. Reeve the pointed ends of the top-pendants up through these, and clinch them over the lower caps, between the doublings of the masts— having the *lower* blocks of the pendant-tackles hooked to the thimble in the extremity of each top-pendant, (hanging about two or three feet below the yard,) so that the hauling-part of the fall shall lead *up*, through a leader lashed on the yard in a line with (and to a block at) the lower cap—thence on deck. Pass stout straps with jambing turns around the crown, and the shank of the anchor near the stock, to which hook the fore and main yard purchases respectively. Rack both parts of the fore-topsail tye aloft, and hook the lower block of the halliards to a strap around the shank, and lashed to a point on the upper arm of the stock about one-third up. Now bend a hawser to the crown, (to weigh the anchor in case of parting the straps,) and man the purchases—keeping a strain on the halliards to cant the stock vertically, and keep the arms in a horizontal position. As soon as the anchor is clear of the lighter, drop her astern, and hook on athwartship tackles to straps around the shank. When high enough, haul over—place the score (in the lower end of the stock) in the bight of the trip-stopper,* rouse the palm in on the gunwale, and lower away—resting the after part of the shank on the trip-shores, which are fitted with a stout hinge in the heel, bolted to the ship's side, permitting them to fall out-board, and bear the anchor clear when let go.

Having thus landed the anchor safely in its berth,† (using a fore-and-aft tackle if requisite,) pass the lashings, consisting of (half-worn) inch rope, on both ends, through span-tackles in the water-way, around the shank, at the stock and crown respectively,‡ covering them, at the conclusion, with canvas, which should be protected from the influence of moisture by a coat of paint or coal-tar. Now shift over the purchases on the opposite side, and secure the yards as before, getting in the other anchor in a similar manner; after which, send down the tackles and burtons—square the yards—cut the racking of the tye aloft—re-hook the halliard-block to the bolt in the channels, and unbend the hawsers from the crown of the anchors.

NOTE.—The lately invented anchor (with moveable arms, which are attached to the crown by a bolt is now in frequent use in the merchant service. This possesses many commendable features—among which we may enumerate small stowage; certainty of holding on, (arising from the peculiar position of the fluke in the ground, and the impossibility of fouling the chain;) less danger to the vessel's bottom in driving over in shoal water, than with the ordinary anchor; and last but not least, admitting the possibility of supplying a broken fluke. These are certainly considerations of no little magnitude, and waiving for the present the question of strength, (which appears to be a mooted one among nautical *connoisseurs*,) it may not be invidious to predict a universal adoption of this admirable invention within a limited period.

*This consists of a stout short span of chain secured horizontally by eye-bolts to the side, and is used for the stock to turn upon, and cant the anchor in letting go.

†The position of the sheet-anchors depends upon the peculiar class and construction of the vessel—sometimes in the fore-channels and sometimes in the waist. In reference to the *trim*, this consideration should not be passed over.

‡In the "Cumberland," in addition to the lashings here described, fore and aft ones were used at the stock and crown, from ring-bolts in the fore-channels and waist.

The following "Table showing the size of Cables and Anchors proportioned to the tonnage of vessels," is extracted from the "*Mechanics' Pocket-Book,*" a very useful little work by Charles H. Haswell, Esq., Chief Engineer U. S. Navy.

Tonnage.	Diam. chain in inches.	Wt. Anchor in lbs.	Wt. Chain per fath.
200	$1.\frac{1}{16}$	1176	60
240	$1.\frac{1}{8}$	1400	70
270	$1.\frac{3}{16}$	1456	78
320	$1.\frac{1}{4}$	1680	86
360	$1.\frac{5}{16}$	1904	96
400	$1.\frac{3}{8}$	2072	104
440	$1.\frac{7}{16}$	2240	115
480	$1.\frac{1}{2}$	2408	125
520	$1.\frac{9}{16}$	2800	136
570	$1.\frac{5}{8}$	3360	144
620	$1.\frac{11}{16}$	3920	152
680	$1.\frac{3}{4}$	4200	161
740	$1.\frac{13}{16}$	4480	172
820	$1.\frac{7}{8}$	5600	184
900	$1.\frac{15}{16}$	6720	196
1000	2.	7168	208

N. B. In all the vessels of the naval service, from 74's to sloops of the 3d class, four anchors (two bowers and two sheets) are allowed, varying from 10,500 to 2,800 lbs. each, according to the vessel for which they are required.

KEDGES.

These are considered indispensable requisites in men-of-war, and a glance at the "Outfit Tables" enables us to state the number allowed to each class of vessels in the naval service.

Class.	No	Weight.
74's, . .	5	1200 to 500
Frigates,	4	900 to 400
Sloops, .	3	600 to 300

NOTE In addition to these, the same number of *boat* anchors is allowed to each vessel, together with two grapnells.

In the equipment of merchant vessels, on the contrary, kedges are a species of "scarce article." This circumstance, of course, places one in a sort of *non-plus* condition with reference to a statement of allowances. The following method of "puddening" is that usually practised in the navy. For the *number* of *ends*, and the *size* of stuff necessary, measure the exact *thickness* of the ring, when ready for puddening, (i. e. when parcelled and woolded over,) and count the strands.* The "ends" are then cut in length, *three* times the *diameter* of the ring. Put a couple or three thicknesses of well-tarred canvas around the ring, and marl them on well—then with a strand of rope well opened, woold them over. Secure the ends to the middle point of the ring with a temporary seizing, and heave them well taut with a tail and heaver *against* the lay of the rope, on one side, and then the other, alternately, two or three times—having a hand to work them around, and to clap on temporary seizings at each heave, until the puddening is as taut as possible. Finally, put on temporary snake seizings at the quarters and ends of the puddening, which latter should be cut square, and fagged well out. The kedges are then stowed in the main and mizen chains, where they are lashed to the lower dead-eyes of the rigging, *inside*.

NOTE.—Where it becomes necessary to anchor among rocks, and there is danger of losing the kedge, the following simple precaution will often be found advantageous: In lieu of bending the hawser to the ring in the usual way, take it around the *crown* with a *buoy-rope bend*, and *stop* it to the ring with a rope-yarn seizing. If now the kedge should get foul, rouse on the hawser, which will part the seizing, and bring the kedge up *crown* first. So long as the boat rides head to wind or tide, the hawser will lead in the direction of the shank, and the effect will be the same as if it were bent to the ring. If, on the contrary, there is no wind or tide, there will be no strain to part the stop, and the result will be identical.

HAWSERS.

While treating of "Ground Tackling," it seems not out of place to refer to the subject of hawsers; more, however, with the view to introduce one or two simple points in seamanship, than to swell the work with "Tables of Allowance"—which "may be seen at all hours" in the Equipment Circulars of the Naval Bureaux.

TO CROWN A HAWSER.—The following rule will be found quite practicable in the performance of this operation: Put a stout whipping on the hawser, a sufficient distance from the end to allow for crowning. Unlay the strands to the whipping, and lay the three *inside* or heart strands up together. Then form the crown with the three *outside* ones, taking them above, and covering the remaining three, which, with the heart strands, should be whipped, and cut off even. Lastly, worm the ends of the crowning strands back into the lay of the hawser, and clap stout smooth seizings close up to the crown, and at the extremity of the worming.

TO SPLICE A HAWSER, OR SHROUD.—At sea, and in the absence of a proper rope-walk, this is often necessary: Unlay the rope sufficient to form the splice, and cut the strands of unequal lengths, (as represented in the following figure,) in order to prevent the splices from coming together when laid up. Take three strands of the

*In any case the rule is invariable; for if the rope used is large, the strands and consequently the number of "ends" will be less; and vice versa.

same-sized rope, (of the requisite length,) and "long-splice" them separately into the strands of the hawser, or shroud. Then secure three large snatch-blocks* at some convenient place, equidistant from each other, and to form a triangle; seizing the strands to the hooks, to which also heavers are lashed, as illustrated below. Have a large block of the same nature, with a heaver attached, secured to the extremity of the shroud in like manner, and bring the whole taut by means of ropes rove through the sheeves of the blocks—observing that the strands bear an equal strain. Put in a ropemaker's *top*, (having a hand to keep it in its place;) man the heavers, and heave round steadily, and on all the strands alike, in a direction *contrary* to the lay of the rope, which is formed at the same time, by turning the heaver on the shroud in the *opposite* way. When sufficient has been laid up, whip the end *abaft* the "top" before taking it out—remove the blocks, &c., &c. This is called a "ropemaker's splice."

No. 29

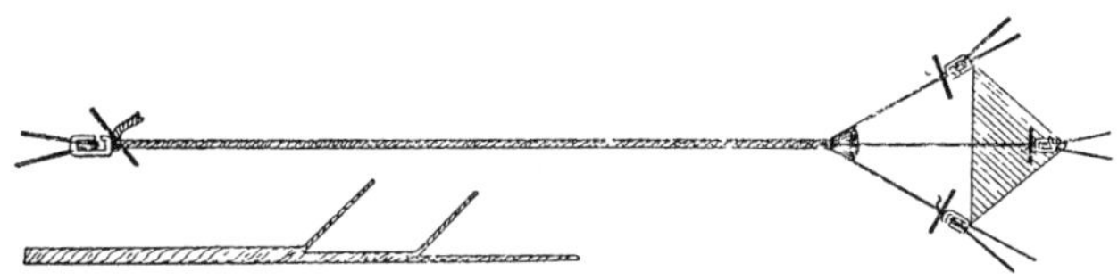

NOTE.—Instead of long-splicing the strands together, the yarns are often singled and *knotted*.

MESSENGERS.

These are now confined exclusively to the naval service, having long since disappeared from merchant vessels, on account of the introduction of the windlass, to which the cable is immediately brought in heaving up. In the fitting, various methods are practised—the opinions as to whether a toggle or lashing is best, being divided among the "sages" of the profession, although all agree as to the kind of rope required, viz., *hawser-laid*, which (by being of a closer lay) is better calculated to resist the action of moisture, and to make the nippers hold on. In cutting the messenger, allow three times the distance from the after part of the capstan to the manger-board, which will afford sufficient for the eyes and turns, of which latter there are *three*. The method of a toggle, though possessing many advantages in point of expedition in shifting and dipping, is nevertheless ill adapted to heaving up, on account of causing the turns to ride, and injuring the welts of the capstan. Under these circumstances, a preference may be said to exist in favor of the lashing. This is passed on *both* ends through the eyes, (which are about two fathoms apart, and short-spliced,) figure-of-eight fashion and seized with a rope-yarn stop to each other temporarily. A reference to tht Rigging Tables of the naval service, gives us the following estimate sizes—there being *two* allowed to each vessel:

For	74's	15 to 14 inch.
"	Frigates,	13 to 12 "
"	Sloops,	11 to 10 "

*The object in using *snatch*-blocks, is on account of the advantages they afford by having a *swivelled* hook.

In passing the messenger, three turns (one above the other) are taken around the welts of the capstan, *with the sun,* if for the *starboard* cable, and *vice versa*—observing that the *under* part is that by which the cable is brought in: Then pass the end forward on the opposite side of the capstan, and lash the eye to that in the other, which is led under the bowsprit, and before the manger-roller. After weighing one anchor, and in order to use the messenger for the other cable, it becomes neceessary to "dip." This is effected by simply casting off the lashing, and passing the end of the *upper* part *down,* or that of the *lower* one *up* through the turns which should be slacked for the purpose) around the capstan. Finally, render the parts taut around the welts, and lash the eyes afresh.

NIPPERS.

These are an essential accompaniment to the messenger and are usually formed of yarns, which are marled together after the manner of a selvagee—having a whipping at each end, and about four and a half fathoms in length. In weighing the anchor, the nippers are successively applied, by first taking a round-turn around the cable, and a round-turn around the messenger, worming the ends into each respectively, and working in all cases from *forward, aft.** In addition to these, selvagees and heavers are hove taut over both messenger and cable, particularly where the latter comes in muddy, and the heave is heavy; and often under these circumstances a bucket of coarse beach-sand is found serviceable in helping the nippers to "bite."

BUOYS.

The use of these articles is confined mostly to the naval service, since the generality of merchantmen are more frequently moored to wharves and docks, than anchored in the stream. Under these circumstances, the latter are ridded of the care which the ordinary buoy naturally requires; and in the event of needing a mark to "betray where the anchor is hidden," a spare dock-fender of white pine, having a hole in one end for the rope, is made to serve the purpose. In men-of-war, the common *nut-buoy* (which derives its name from the close resemblance it bears to a pecann-nut) is in most general use. Of late many improvements have been made in this—and we may enumerate in particular, that of making it of *sheet-iron,* which, in addition to saving the amount of labor and gear incident to the old-fashioned slings† has the advantage of *stopping its own leaks,* by the creation of *rust,* which arises from a constant exposure of the iron to the corrosive action of salt water. In this age of *gutta percha,* however, even this plan might be modified, (for stowage and economy at least,) by having the buoy of *India-rubber,* and protected from chafe by a netting of rope. In cases of shipwreck, it

*The effect of this method is, to preclude the possibility of the turns coming together and jambing, which manifestly can never take place, since each end of the nipper is drawn in contrary directions by the cable and messenger—while *both* ends are clear for taking off.

†These are cut in length *nine* times that of the buoy, which allows sufficient for the straps and slings. These, in the fitting, should be well stretched, and protected by worming and service.

might also be employed as a *life-preserver*. The *size* of the buoy is usually about *one-third* that of the cable; or, since this rule applies more particularly to hemp-cables, of strength sufficient to weigh the anchor by itself, in the event of its being necessary. The *length*, of course, depends upon the depth of water. In ordinary cases this is attached to the crown by what it technically termed a buoy-rope bend, which is formed by taking a half-hitch over each arm, and stopping the end along the shank. Many suggestions have been made by nautical writers to avoid the use of the buoy-rope here described, which becomes, on account of its size, a troublesome and unwieldy fixture. As yet, however, none have been found to fulfil the desired end; and that recommended by Lieut. Totten, in his *Naval Text-Book*, has the greatest of all objections, viz. *impracticability*. This arises from the parts of the small rope (rove through the strap on the crown) becoming *hawser-laid*, and precluding the possibility of reeving or unreeving, for the passage of the "stout rope of the required size," which that author suggests, "to avoid the inconvenience" of "ordinary cases."

STOPPERS.

These are used for many purposes, but most generally for that of securing the cables in the tiers and on deck; and as such, distinguished by names as various as their uses. Improvements, however, (as in the other branches of the profession,) have been made in stoppers, and those used at the bitts (in the *naval* service at least) are now almost universally fitted of chain, with a devil's-claw, large enough to receive one of the links of the cable, over which it is placed, and retained by a small iron-pin, running through both parts of the claw. In the other extremity a slip-hook and ring is attached, by which it is secured to the stopper-bolts of the deck. Its length is about four feet and a half, and in size depends upon the class of vessel for which it is required. This is designed to supersede the ordinary *deck-stopper*, which is usually fitted with a double-wall and crown-knot in one end, and in the other a hook and thimble, securing it to the ring-bolts in the deck. This, which is formed of hawser-laid rope, about a fathom and a half long, is applied to the cable either before or abaft the bitts (as may be necessary) by means of a laniard passed around both cable and stopper, (from *forward, aft*,) usually in six turns, and the end, which is brought up between both, wormed forward, and seized to the former. With a *hemp*-cable, the laniard is passed *against* the sun, and wormed into the lay.

The universal adoption of the windlass in merchantmen, (to which we have already had occasion to refer,) necessarily obviates many of these appendages, which are regarded in the navy as essential requisites. In lieu of the stopper described above, a simple *bitt-stopper* is employed to nip the cable on the bight forward of all. This usually consists of four or five fathoms of rope, tailed out like a nipper at one end, and fitted with a clinch knot in the other, by which it is secured through a hole in the foremost samson-knee.* These, with the exception of *belaying*-stoppers, (which are so simple as to render a description unnecessary,) are about all that con-

*In merchantmen, stoppers *abaft* the bitts are seldom used, on account of the lockers being so near the windlass. In ordinary cases, a small deck-jigger is clapped on, and hauled taut; and in *extra*-ordinary ones, the cable is weather-bitted, i. e. by taking an additional turn around the welts of the windlass.

cern the merchant service. In the navy, on the contrary, hatch and tier stoppers are in frequent use—although the lately-invented *compressor** is found to obviate greatly the old and almost numberless phalanx of "cable-gear," which even now lingers as a vestige of vanished days.

NOTE.—On board of the "Cumberland," in coming to anchor, it was always the practice to let the anchors take the chains from the lockers—using the compressors to stopper when required. This plan invariably succeeded, and saved a great deal of trouble. In mooring, the *lee* cable (for veering) was unbitted.

LOG-LINE AND GLASS.

To attempt an enumeration of the disasters which have resulted from the neglect and imperfection of this simple instrument, would occupy more space than the already prescribed limits of the work will allow. Yet the repetition of a few recognised principles seems not inappropriate, while referring to the equipment of the log-line and glass. The late Captain Glascock, who lays more stress than is usually bestowed upon the subject by nautical writers, aptly expresses himself thus: "The 'dead reckoning' is seldom unchecked by other data, and people who perform long voyages, and trust to astronomical observations, pay but little regard to the 'run' given by the log. The naval astronomer, with his chronometer, sextant, and 'Nautical Almanac,' and with the sun, moon, and stars unobscured, feels for a time independent of the log; but of what avail is astronomical knowledge, if, towards the termination of a passage, when pushing for a port, the heavenly bodies become invisible to the human eye? This is a circumstance by no means unusual in temperate climates for days and nights together, particularly on approaching the *land*. At periods like these, even the astronomical navigator must lay aside the watch and sextant for the ruder instruments of the lead and log. Let him therefore bear in mind that an incorrect return of the ship's periodical 'run' may be the sole and immediate cause of her total loss." With regard to the merchant service, these observations seem directly suited, although it is by no means intended to place the practice of navigation in the navy beyond the pale of their application to it also. For though the commanding officer himself, and the master (under whose particular province the log-line and glass fall) may inspect their adjustments as frequently as may be essential to their correctness, yet the carelessness of those who *heave the log*, more than counterbalances the errors which would result from an *incorrect log, properly* hove.

In the fitting, an observance of the following points will be found to render the log sufficiently accurate for practical purposes. *Wet* the line, which should be well stretched, before cutting, and allow half the length of the ship's keel, for the "stray," which point, mark with a white rag. Then "divide the knots into portions of forty-seven feet three inches, to the twenty-eight-second glass,"—and mark this distance (correctly measured) with brass or copper tacks in one of the *midship* deck-planks, for further reference; observing every third inspection of the line,

*This (of which in the Cumberland there were two to each bower, one on the berth-deck, and the other in the hold) consists of an iron elbow, bolted in one end to the under side of the beam just forward of the chain-pipe. The other has usually a small luff-tackle attached, which is used to nip the compressor taut around the chain.

which should be done once a week, to shift it end for end. The reel, in order to turn with ease, were best, if fitted with friction rollers—having the cheeks about six inches in diameter, and separated by a space of about eighteen inches. In *timing* the glass, (which should be kept in some convenient place free from the influence of *moisture*,) instead of using an ordinary watch, refer to the second-hand of the chronometer.

In *heaving the log*, (which is quite as important as any feature in the equipment,) a portion of the "stray," sufficient to carry the chip clear of the eddy formed by the wake, should be carefully *flaked* (*not coiled*) in the hand, and thrown over the *stern*—the person so doing standing on the lee side abaft, observing that the sails are full, and the ship either "by" or at her course, and directing the assistant to hold the reel well up, careful that no impediment is offered to the free progress of the line. When the first point of reckoning, or the mark of the stray, passes through the officer's hand, he gives the order, "*Turn!*" which the glass-holder repeats, "inverting the glass by a timely turn of the wrist," and watching the moment the last grains of sand fall through, he cries "*Stop!*" The line being immediately stopped, the mark nearest the officer's hand shows the number taken from the reel—estimated in fathoms, which, together with the knots, shows the distance run by the ship in the elapsed time. A calculation of the *average* velocity, with an allowance for the heave of the sea, the steadiness of the wind, and the manner of steering, (during the interval of the previous cast of the log,) is then made by the *watch officer*, and inserted opposite the course. These are all points of the utmost importance, and should have a due share of consideration, as affecting materially the position, progress, and safety of the ship. If it be true in morals, that "charity hideth a multitude of sins," it is not less certain in the nautical code, that correct navigation is the *only* plaster for *bad* seamanship.

LEAD-LINES.

In tracing the various improvements which have been made from time to time in the nautical profession, it is not a little strange that the lead-*line* has been permitted to remain thus long in *statu quo*, with the same rude marks which it probably had in the offset of its use. Yet such is the fact, notwithstanding the number and abilities of professional writers, who have, one by one, handed down, with the care of a scriptural translation, these relics of primitive seamanship.

Some years ago, a "deep sea" lead was invented, which sought to remedy the evils of continual error, and the inconvenience resulting from the *antique* system of marking still practised. This consisted of an instrument of cylindrical shape, containing a small glass tube, having a hole in the lower extremity, protected by a delicate spring, which (acted upon by the upward pressure of the water) permitted a quantity, proportioned to the depth, to rise in the tube as the instrument descended below the surface, which distance was duly indicated by means of a graduated side scale. This contrivance, however, far from succeeded in answering the conditions sought for by its ingenious inventor, on account of the delicacy of the spring, and the corrosive influences of the element to which it was exposed. Under these circumstances, we are forced back to first principles, and obliged to retain the marks on the line, in order to correct the very errors which the instrument was designed to avoid.

With due deference to the *age* of the present method, the following system is offered as one, at least, which has the advantage of revealing the "marks and deeps" without regard to the "*ocular* demonstration" of the existing order.

The *hand lead-line* (which is seldom over thirty fathoms in length) may be marked thus:

At 2	fathoms	2 beads.*
" 3	"	3 "
" 5	"	leather.
" 7	"	1 knot.
" 10	"	1 bead.
" 13	"	2 knots.
" 15	"	rag.
" 20	"	toggle.
" 25	"	2 beads.
" 30	"	3 "
" 35	"	short splice.

In this system no two marks are the same, except those at 2 and 25, and at 3 and 30, which, from their distance apart, never can be mistaken for one another. From an examination of the old rule, we perceive, on the contrary, that 3 and 13, 5 and 15, 7 and 17, and 25 and 35, are identical; and moreover, consist of *white* and *red* rags, as if *these* were distinguishable colors in the *dark!*

Again, the marks on the line of the *deep-sea* lead may be modified for the better by an adoption similar to the following:

At 15	fathoms†	1 bead.
" 20	"	2 "
" 30	"	3 "
" 40	"	1 knot.
" 50	"	2 "
" 60	"	3 "
" 70	"	rag.
" 80	"	leather.
" 90	"	splice.
" 100	"	toggle.

The advantage of this system is, that no two marks are the same, which is far from being the case with the deep-sea line now in use.

In order, however, that the reader may compare the old system of marking with that suggested in the preceding paragraph, the following extract is added from Totten's "*Naval Text-Book.*"

*These consist of small broad flat rings of *lignum-vitae* secured to the line by a wooden pin in each, running through and through—the beads being separated from each other by a space of two or three inches.

†The length of this line, and its application to great depths, renders a mark before 15 unnecessary.

"Both the hand and deep-sea lines are marked as follows:

At	2	fathoms	from	the	lead,	with 2 strips of leather.
"	3	"	"	"	"	with 3 strips of leather.
"	5	"	"	"	"	with a white rag.
"	7	"	"	"	"	with a red rag.
"	10	"	"	"	"	with leather, having a hole in it.
"	13	"	"	"	"	as at 3.
"	15	"	"	"	"	as at 5.
"	17	"	"	"	"	as at 7.
"	20	"	"	"	"	with two knots.
"	25	"	"	"	"	with one knot.
"	30	"	"	"	"	with three knots.
"	35	"	"	"	"	with one knot.
"	40	"	"	"	"	with four knots. And so on.

"In marking the deep-sea line, the first marks before twenty fathoms may be omitted, with the exception of the ten-fathom mark, and if you please, the fifteen, as directed above."

NOTE.—While fitting the lead-lines the *breast-bands* should not be omitted. These are best if made of white cotton canvas, about a fathom in length, three inches wide, and tapering at each end, with eyelets, into which laniards are spliced.

Such articles as chain-hooks, hook-ropes, clear hawse pendants, &c., are so subordinate in their reference to Ground Tackling, and so simple in their construction, that it seems scarcely necessary to swell the subject beyond its present limits with their description.

PLATES

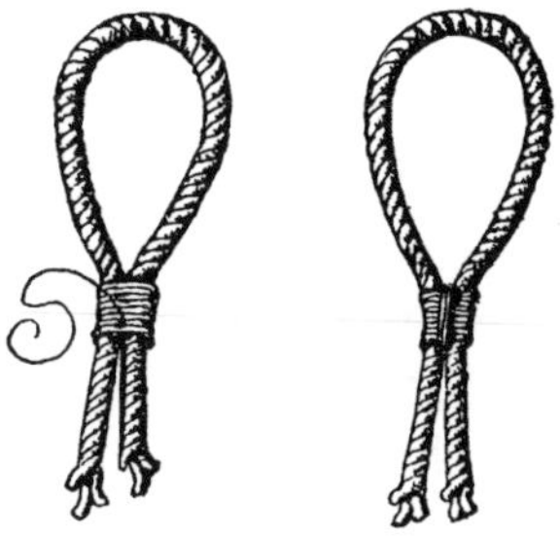

LOWER RIGGING EYE SEIZING
Fig. 1, Page 8

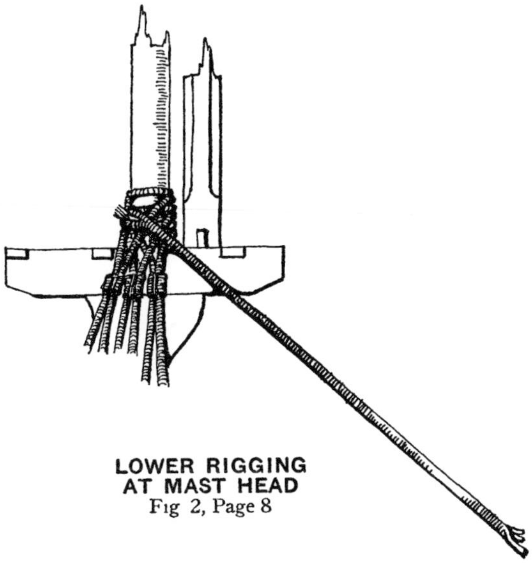

LOWER RIGGING AT MAST HEAD
Fig 2, Page 8

FORE STAYS LASHING EYES
Fig 3, Page 10

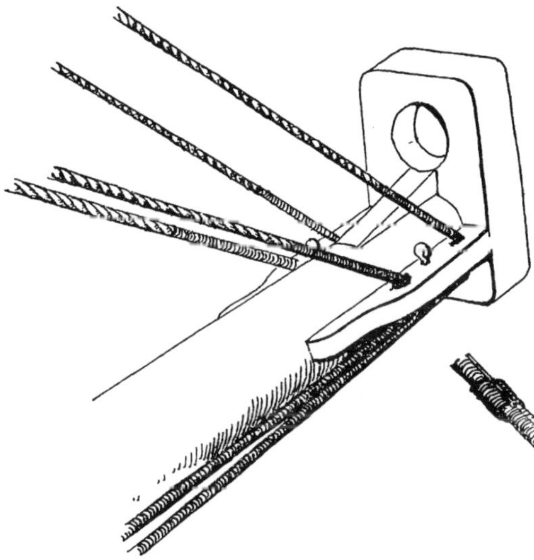

FORE STAYS BEES
Fig 4, Page 10

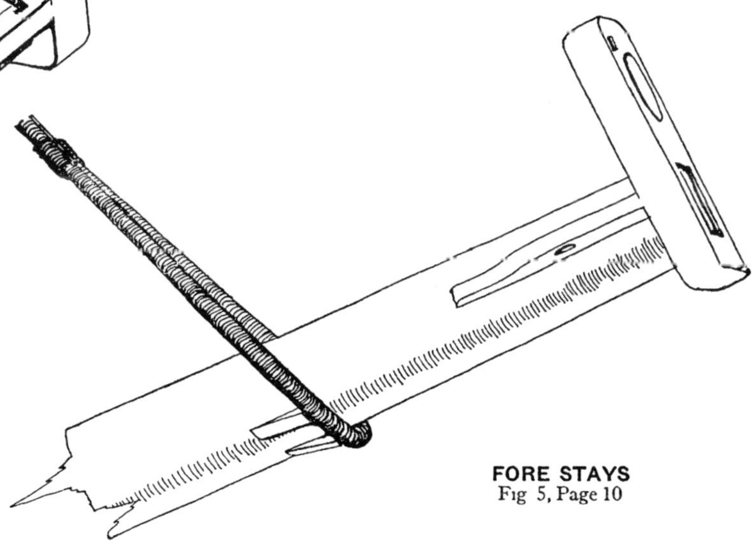

FORE STAYS
Fig 5, Page 10

Plate 1

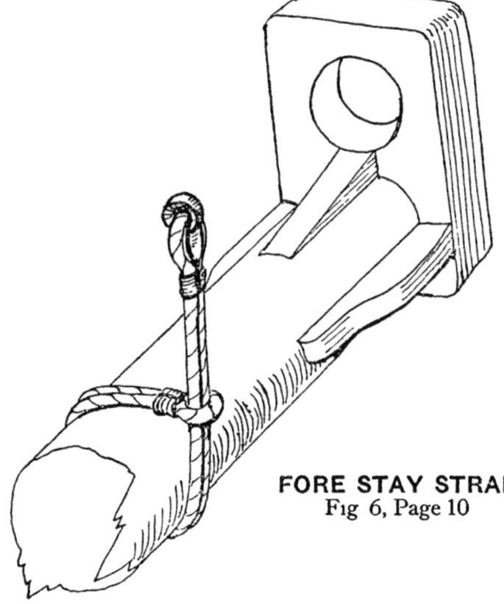

FORE STAY STRAP
Fig 6, Page 10

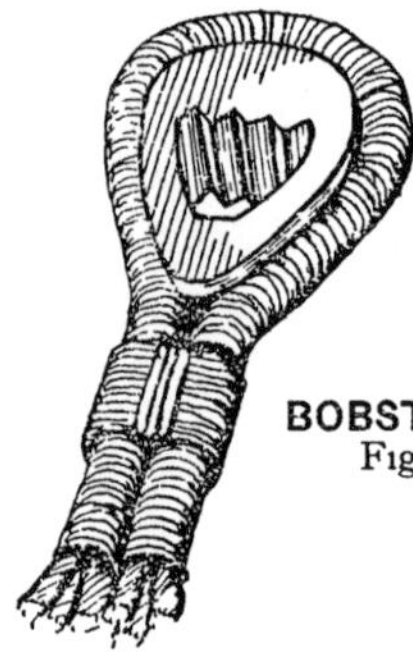

BOBSTAYS HEART
Fig. 7, Page 10

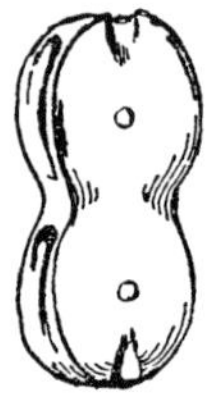

TOPMAST SHROUDS BLOCK
Fig 8, Page 12

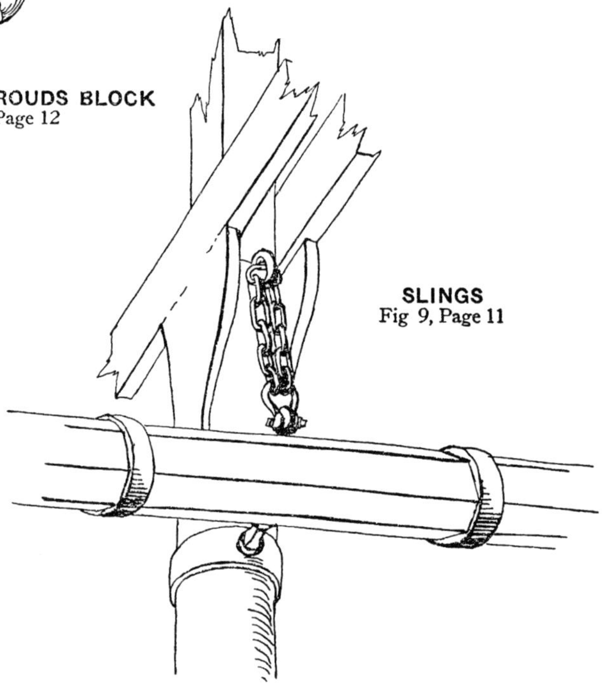

SLINGS
Fig 9, Page 11

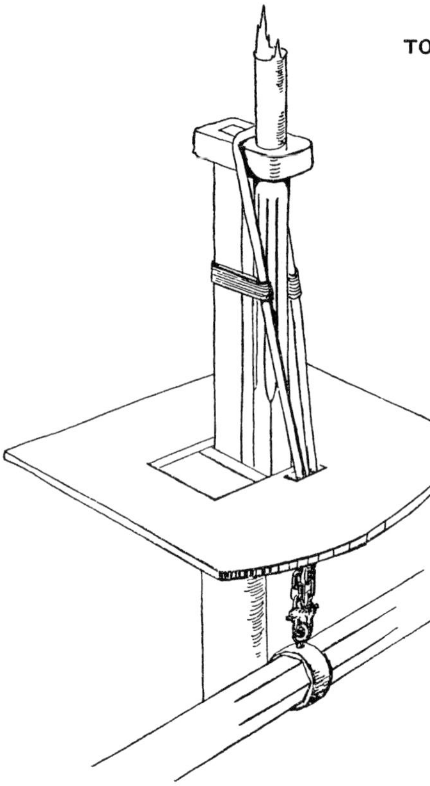

SLINGS
Fig 10, Page 11

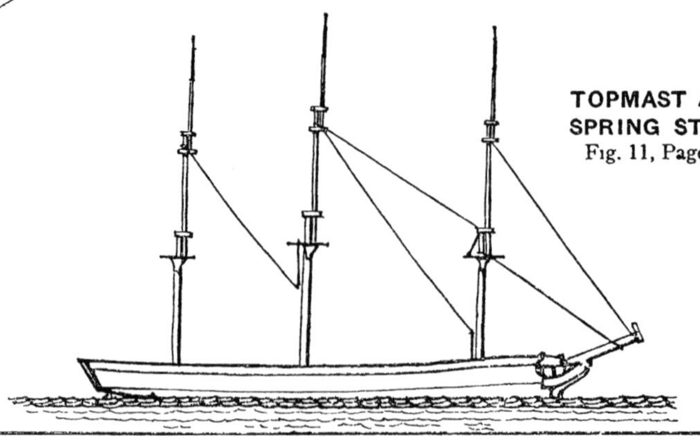

TOPMAST AND SPRING STAYS
Fig. 11, Page 12

JIB STAY
FLEMISH EYE
Fig 12, Page 13

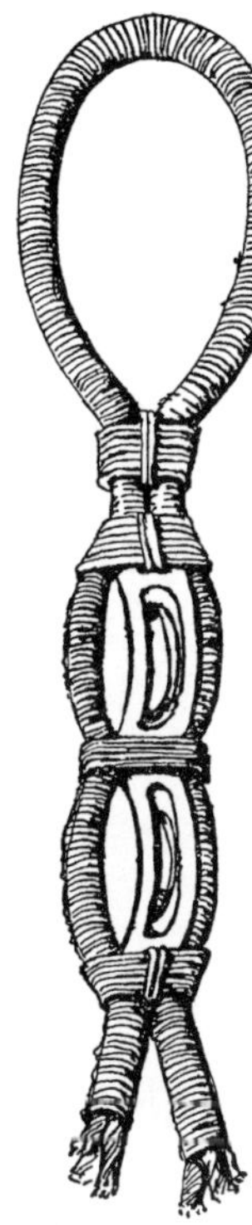

TOPMAST SHROUDS
Fig 14, Page 12

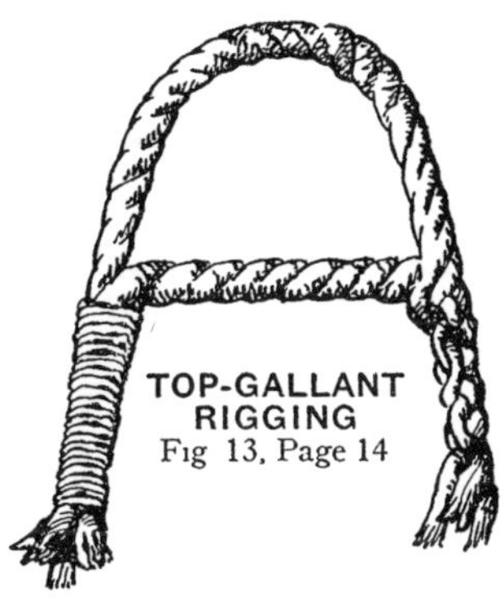

TOP-GALLANT RIGGING
Fig 13, Page 14

TOP-GALLANT RIGGING
CUT SPLICE
Fig 15, Page 13

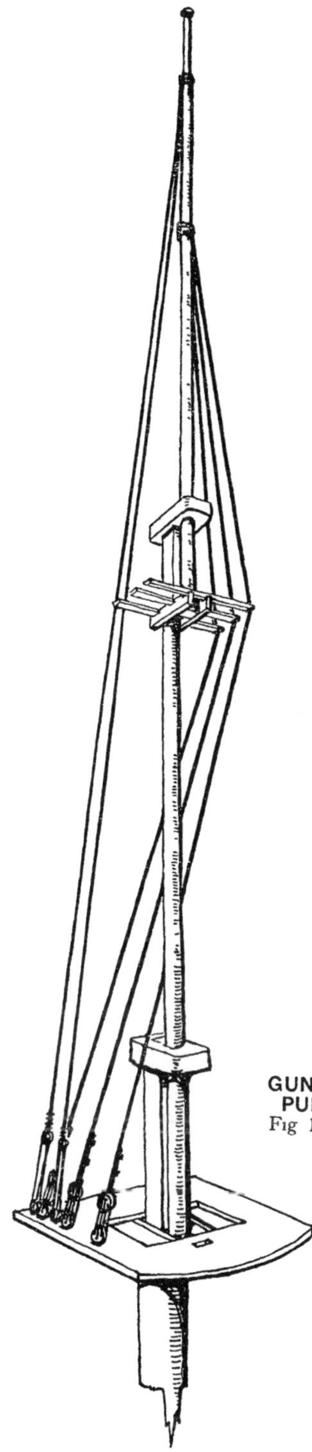

ROYAL RIGGING
Fig 17, Page 14

GUN TACKLE
PURCHASE
Fig 16, Page 14

Plate 3

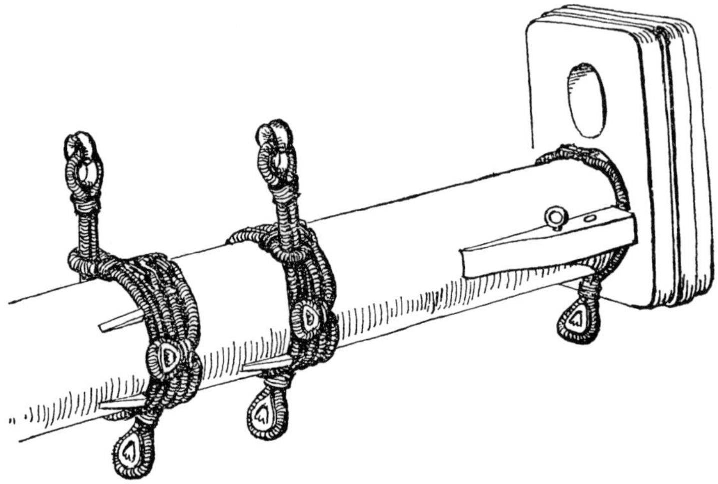

BOWSPRIT
Fig 18, Page 16

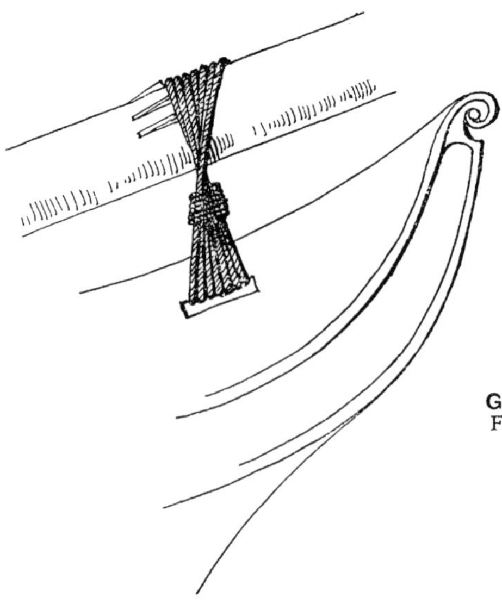

GAMMONING
Fig 19, Page 17

Plate 4

RATLINS
Fig 20, Page 27

CLOVE HITCH IN THE RATLINS
Fig 21, Page 27

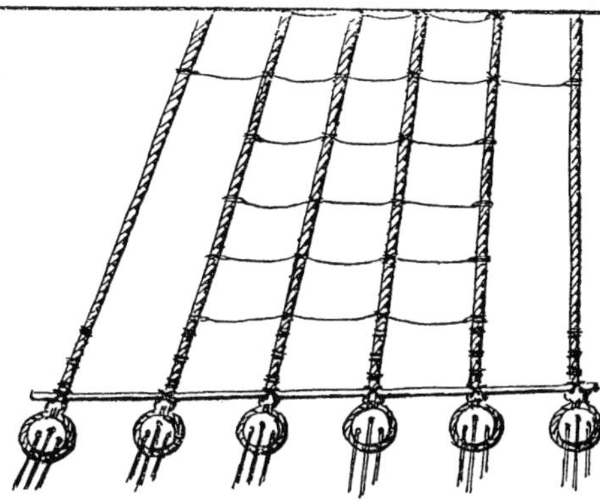

LOWER RIGGING
Fig 22, Page 26

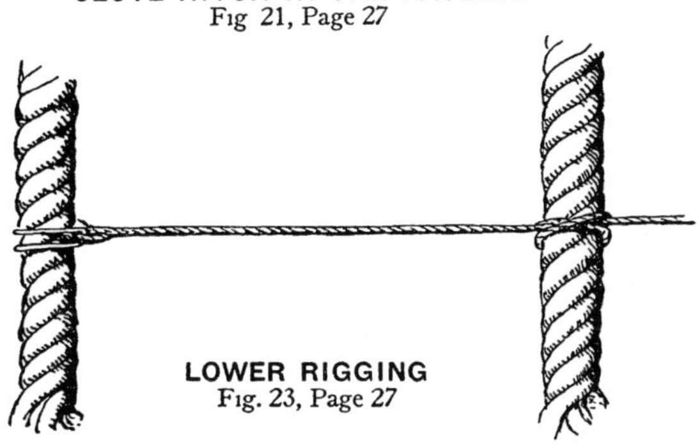

LOWER RIGGING
Fig. 23, Page 27

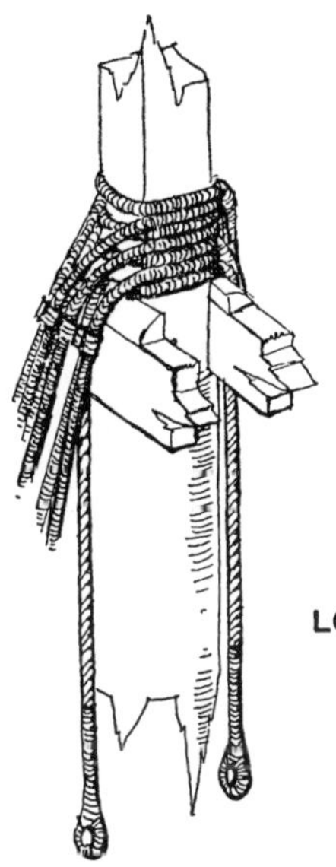

LOWER RIGGING
Fig 24, Page 28

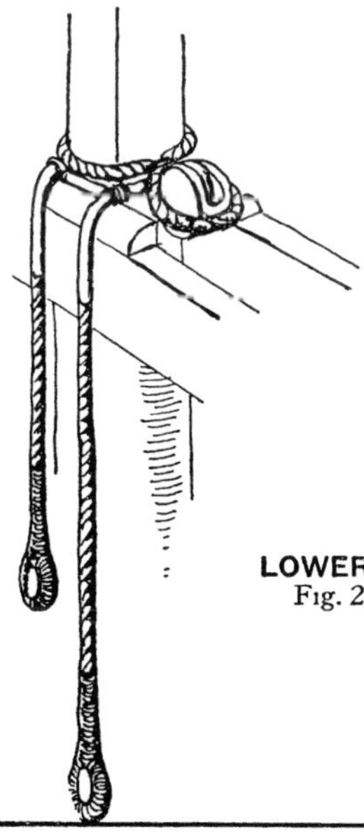

LOWER RIGGING
Fig. 25, Page 21

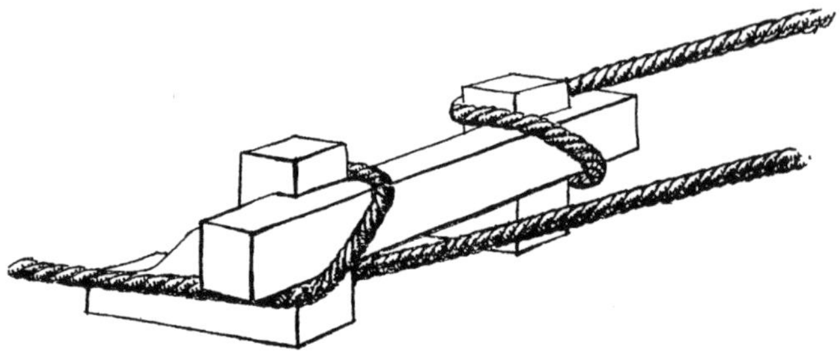

LOWER RIGGING
Fig. 26, Page 28

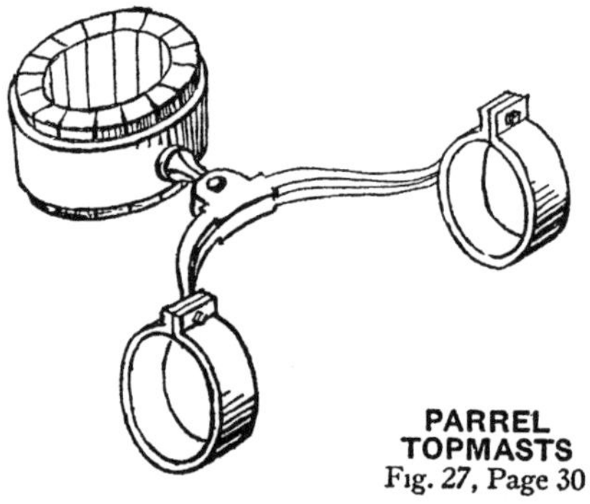

PARREL
TOPMASTS
Fig. 27, Page 30

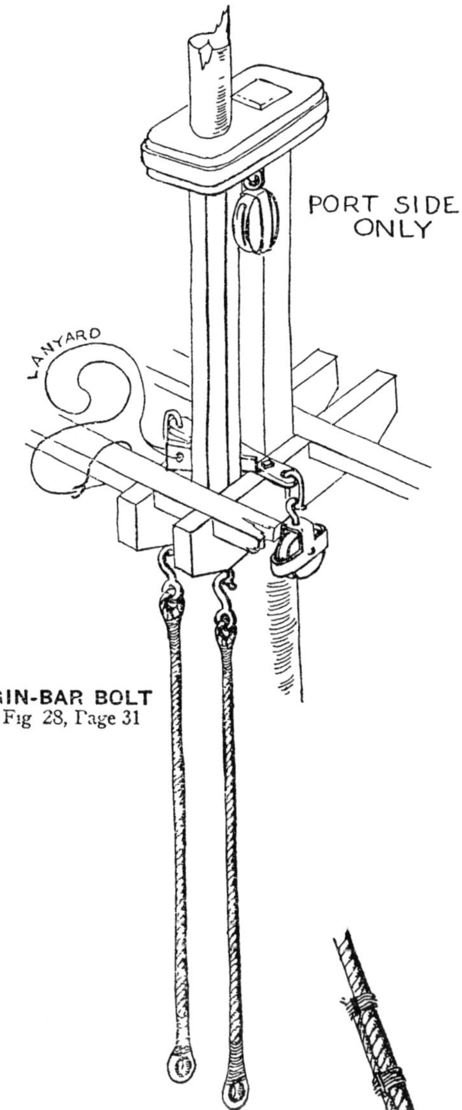

GIN-BAR BOLT
Fig 28, Page 31

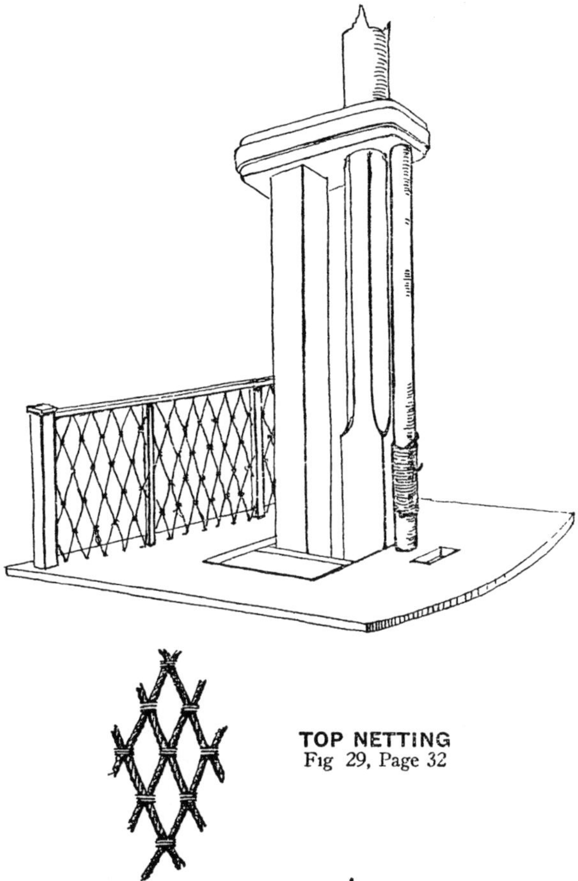

TOP NETTING
Fig 29, Page 32

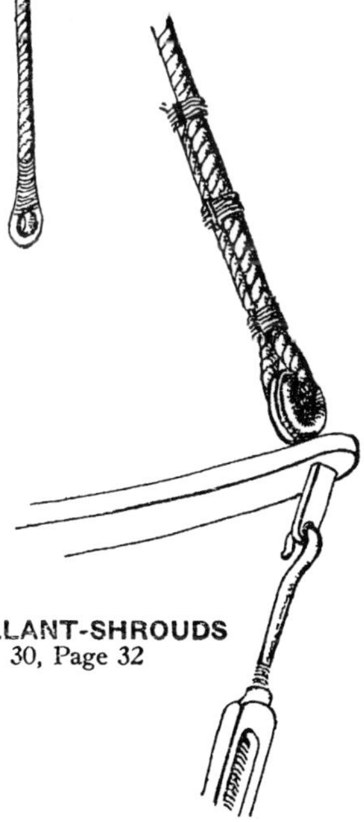

TOP GALLANT-SHROUDS
Fig 30, Page 32

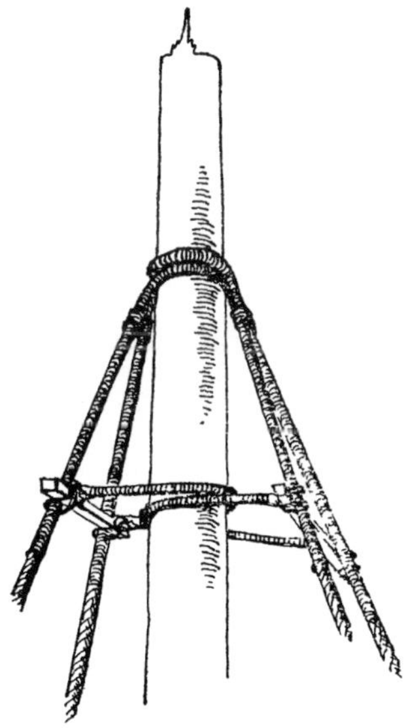

FUTTOCK STAFFS
CAT-HARPEN LEGS
Fig 31, Pages 31, 32

Plate 7

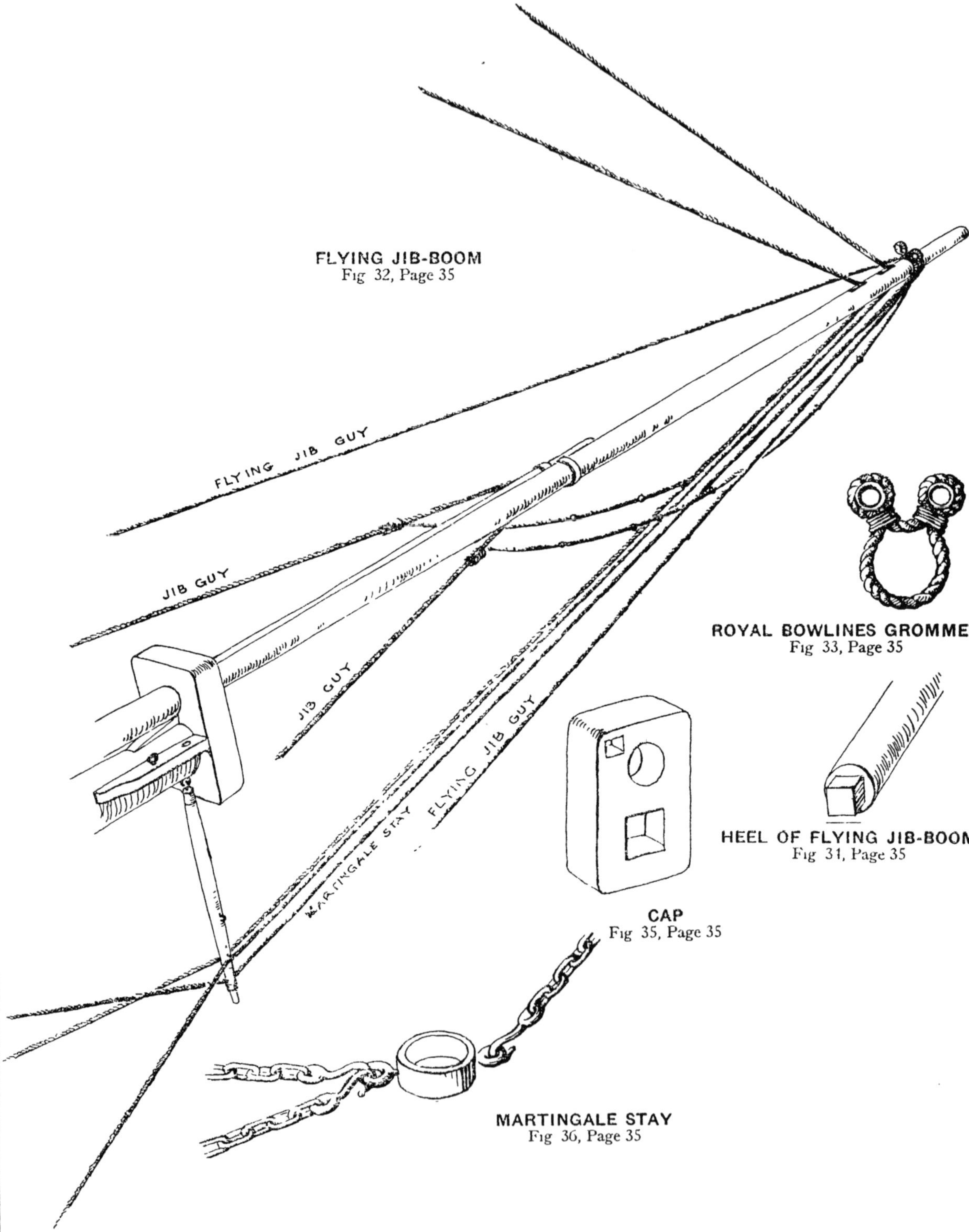

FLYING JIB-BOOM
Fig 32, Page 35

ROYAL BOWLINES GROMMET
Fig 33, Page 35

HEEL OF FLYING JIB-BOOM
Fig 31, Page 35

CAP
Fig 35, Page 35

MARTINGALE STAY
Fig 36, Page 35

Plate 8

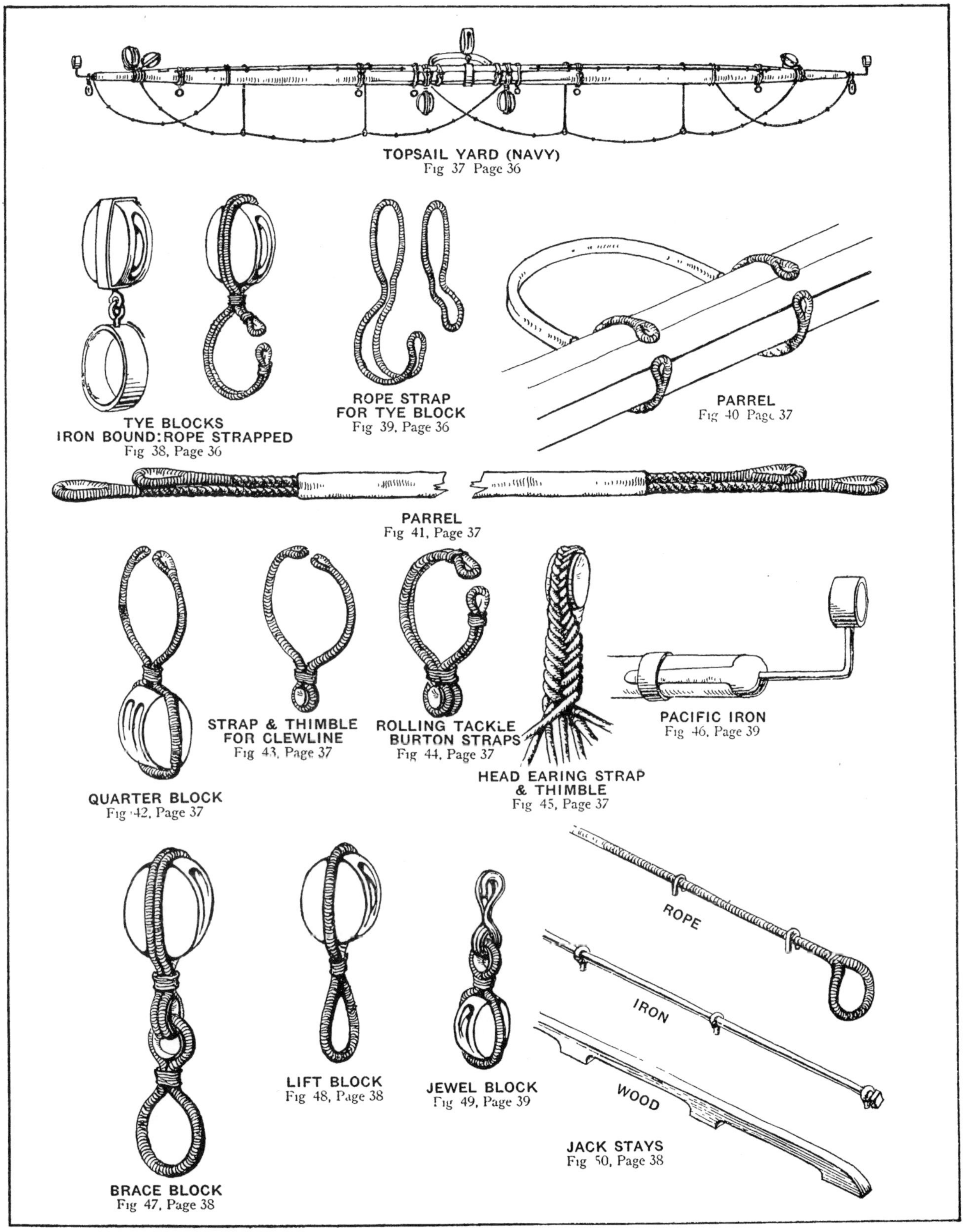
TOPSAIL YARD (NAVY)
Fig 37 Page 36
TYE BLOCKS
IRON BOUND: ROPE STRAPPED
Fig 38, Page 36
ROPE STRAP
FOR TYE BLOCK
Fig 39, Page 36
PARREL
Fig 40 Page 37
PARREL
Fig 41, Page 37
QUARTER BLOCK
Fig 42, Page 37
STRAP & THIMBLE
FOR CLEWLINE
Fig 43, Page 37
ROLLING TACKLE
BURTON STRAPS
Fig 44, Page 37
HEAD EARING STRAP
& THIMBLE
Fig 45, Page 37
PACIFIC IRON
Fig 46, Page 39
BRACE BLOCK
Fig 47, Page 38
LIFT BLOCK
Fig 48, Page 38
JEWEL BLOCK
Fig 49, Page 39
ROPE
IRON
WOOD
JACK STAYS
Fig 50, Page 38

Plate 9

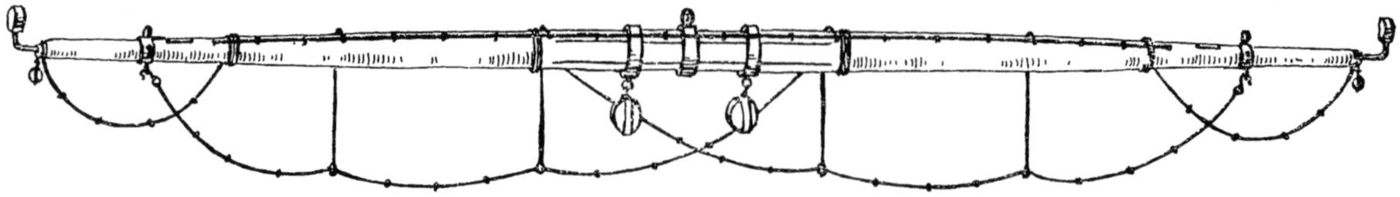

TOPSAIL YARD
Fig 51, Page 36

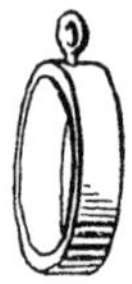

SLING BAND
Fig 52, Page 36

QUARTER BLOCK
Fig 53, Page 36

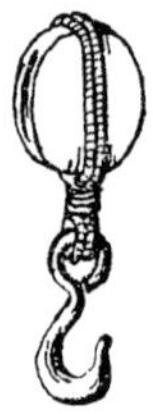

BRACE BLOCK
Fig 54, Page 38

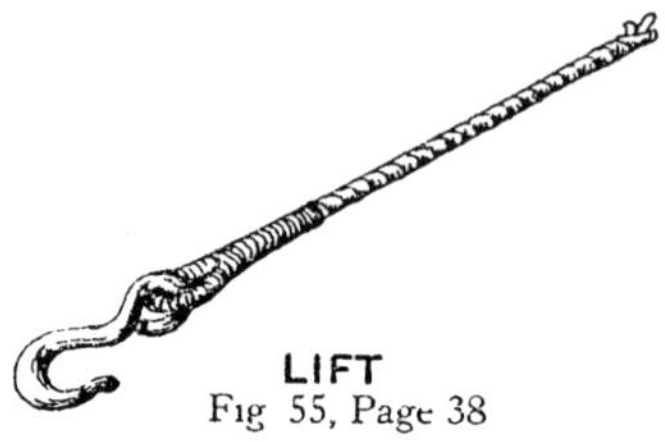

LIFT
Fig 55, Page 38

JEWEL BLOCK
Fig 56, Page 39

FLEMISH HORSE
Fig 57, Page 39

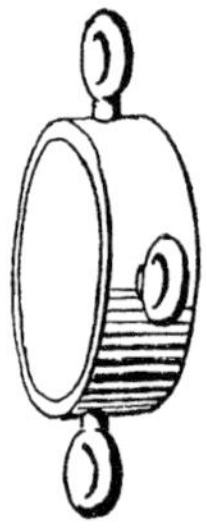

SHOULDER BANDS
Fig 58, Page 38

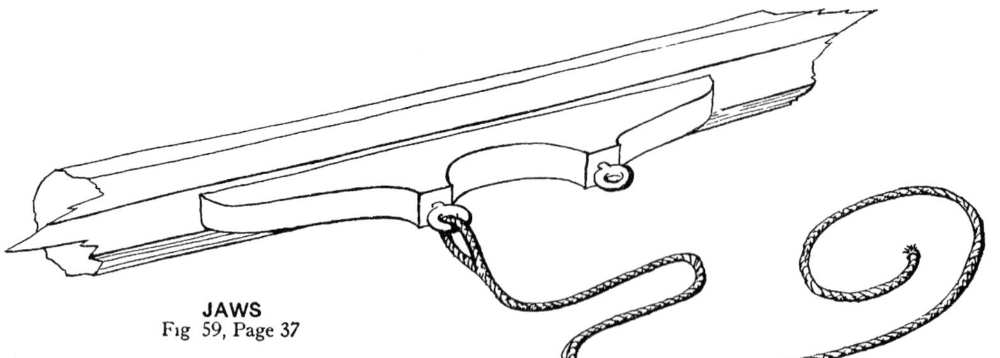

JAWS
Fig 59, Page 37

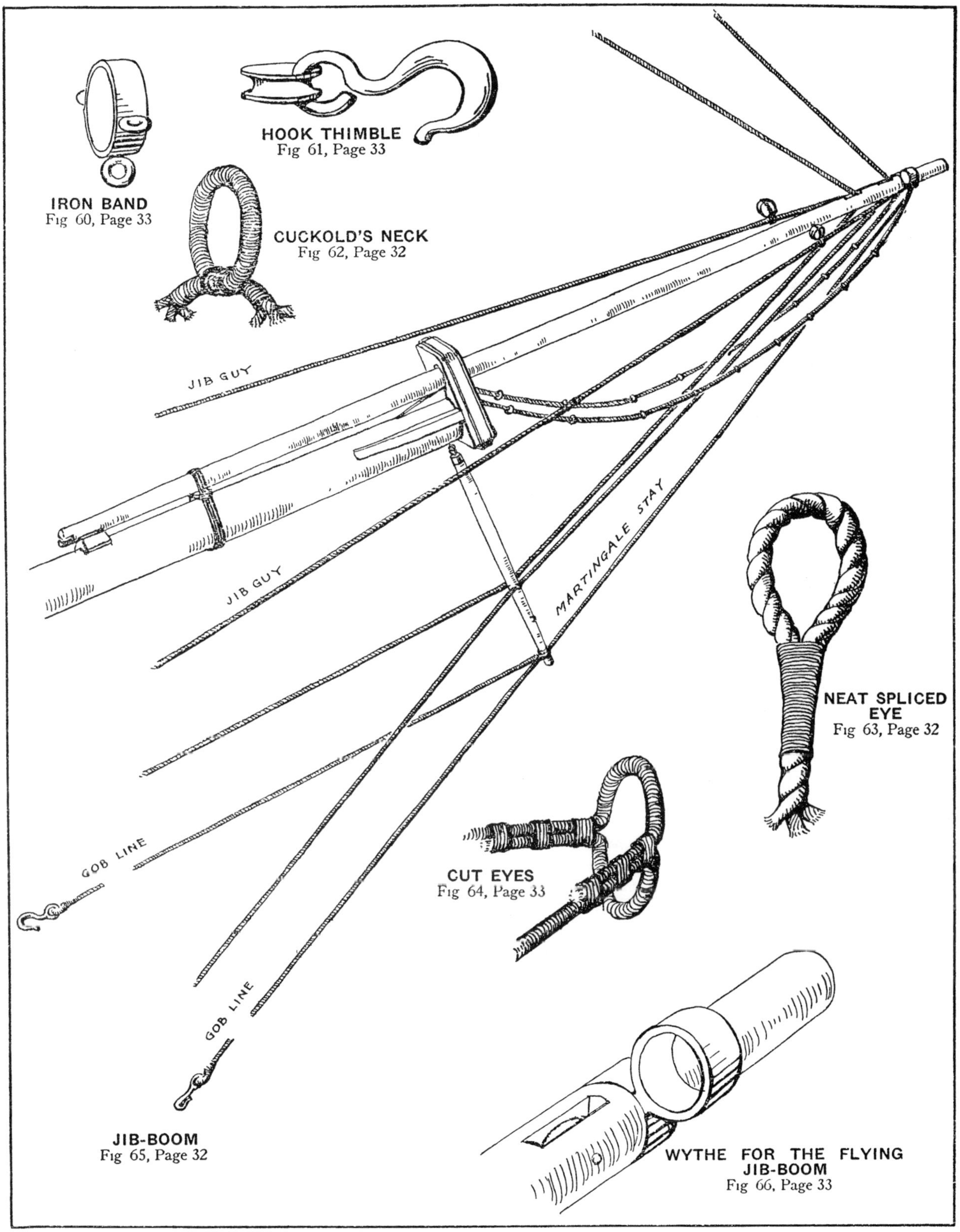
IRON BAND
Fig 60, Page 33
HOOK THIMBLE
Fig 61, Page 33
CUCKOLD'S NECK
Fig 62, Page 32
JIB GUY
JIB GUY
MARTINGALE STAY
NEAT SPLICED EYE
Fig 63, Page 32
GOB LINE
CUT EYES
Fig 64, Page 33
GOB LINE
JIB-BOOM
Fig 65, Page 32
WYTHE FOR THE FLYING JIB-BOOM
Fig 66, Page 33

Plate 11

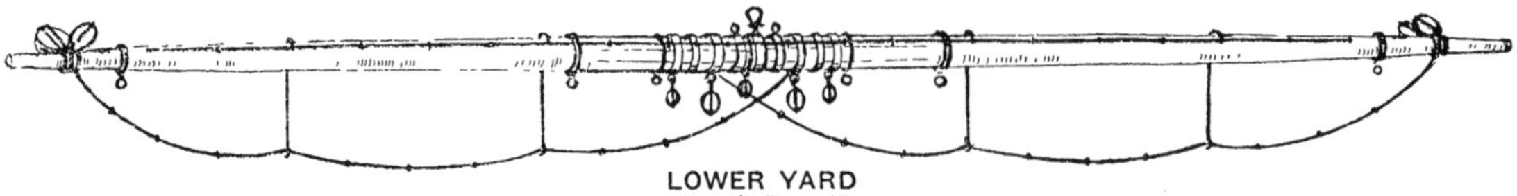

LOWER YARD
Fig 67 Page 40

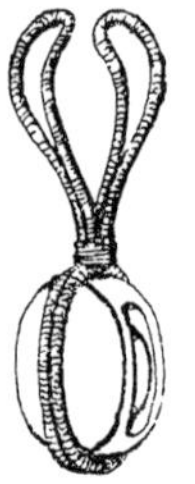

QUARTER BLOCK
Fig 68, Page 41

CLEW GARNET BLOCK
Fig 69, Page 41

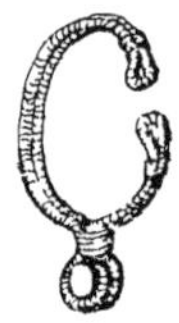

BURTON STRAPS
Fig 70, Page 42

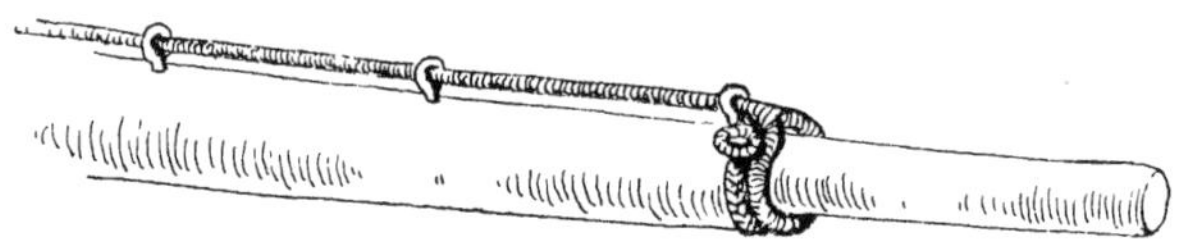

JACK STAY
Fig 71, Page 42

SLAB LINE BLOCK
Fig 72, Page 41

CLEW GARNET STRAP
Fig 73 Page 42

BRACE BLOCKS
Fig 74 Page 42

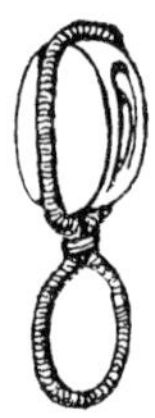

LIFT BLOCK
Fig 75, Page 42

BRACE BLOCK PENDANT
Fig 76, Page 42

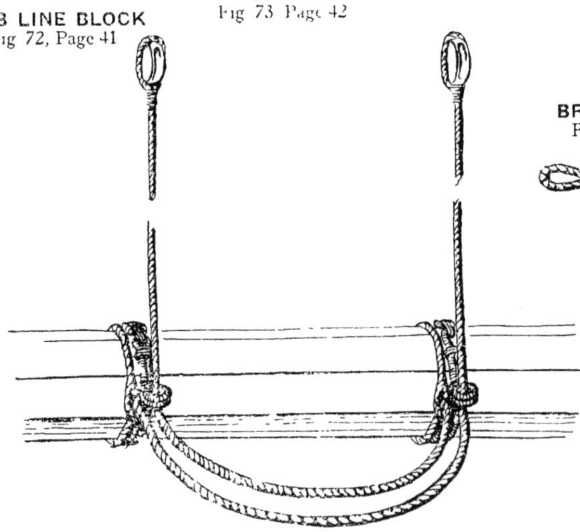

TRUSS STRAPS AND PENDANTS
Fig 77 Page 41

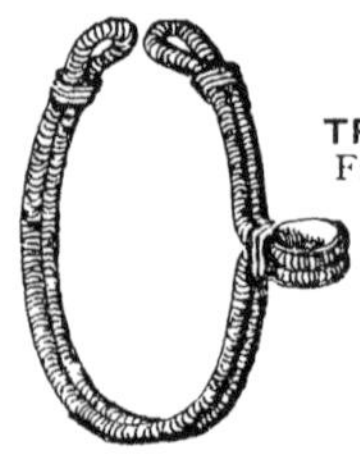

TRUSS STRAP
Fig 78, Page 41

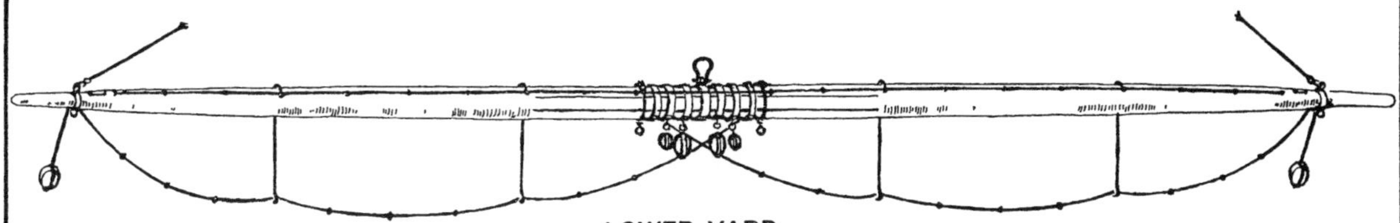

LOWER YARD
Fig 79, Page 40

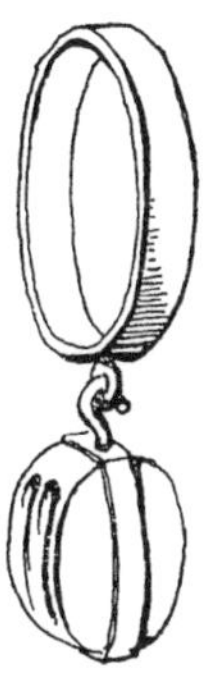

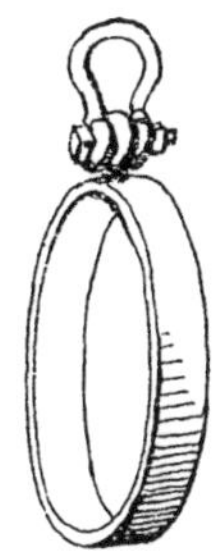

SLING BAND
Fig 83, Page 41

CLEW GARNET STRAP
Fig 82, Page 42

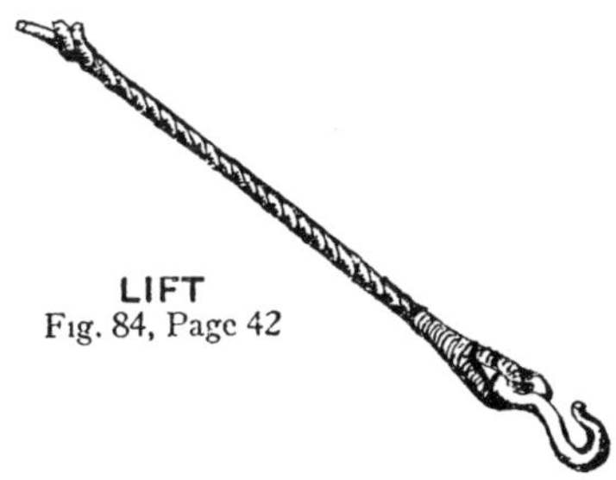

CLEW GARNET
Fig 81, Page 41

QUARTER BLOCK
Fig. 80, Page 41

LIFT
Fig. 84, Page 42

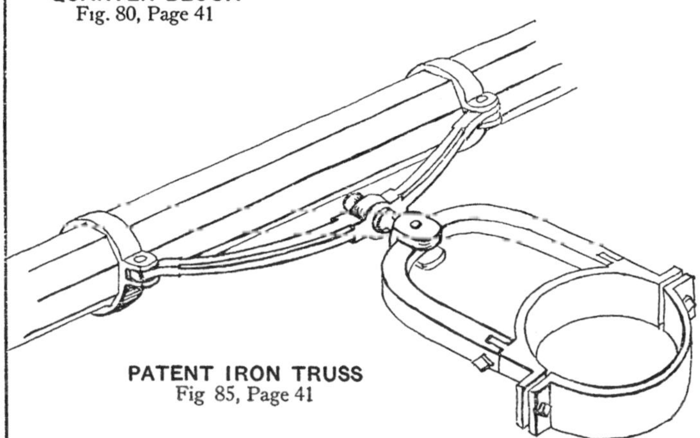

PATENT IRON TRUSS
Fig 85, Page 41

BRACE BLOCK
Fig 86, Page 42

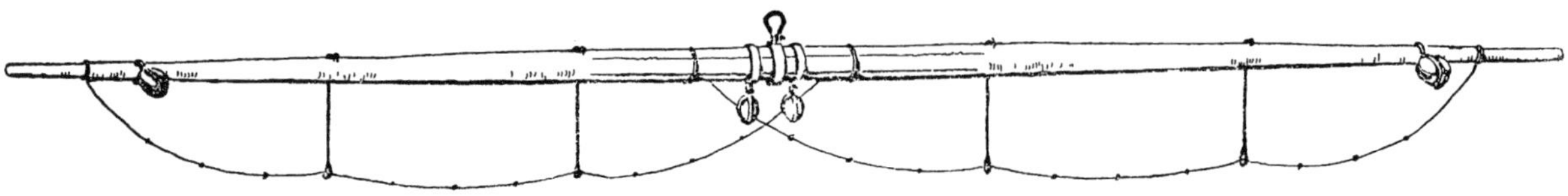

CROSS JACK YARD
Fig 87, Page 43

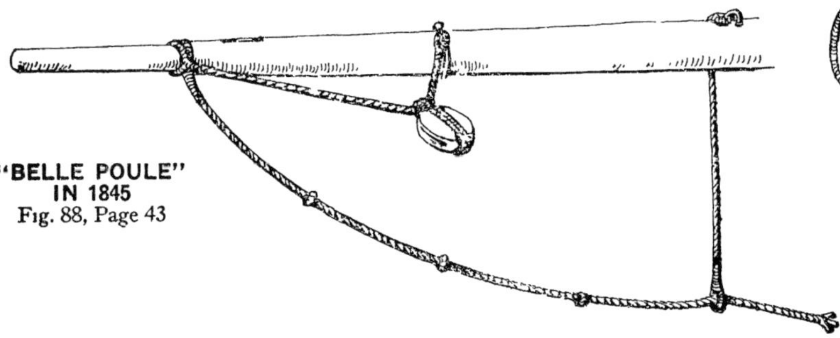

"BELLE POULE"
IN 1845
Fig. 88, Page 43

BRACE STRAPS
Fig 89, Page 43

SLING BAND
CROSS JACK YARD
Fig. 90, Page 43

QUARTER BLOCK
CROSS JACK YARD
Fig 91, Page 43

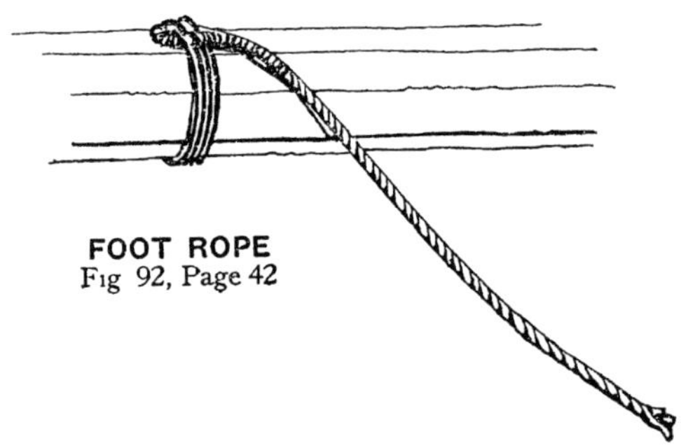

FOOT ROPE
Fig 92, Page 42

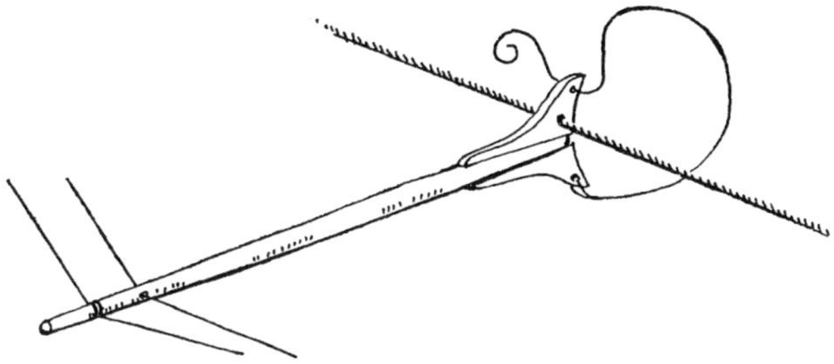

WHISKERS
Fig. 93, Page 43

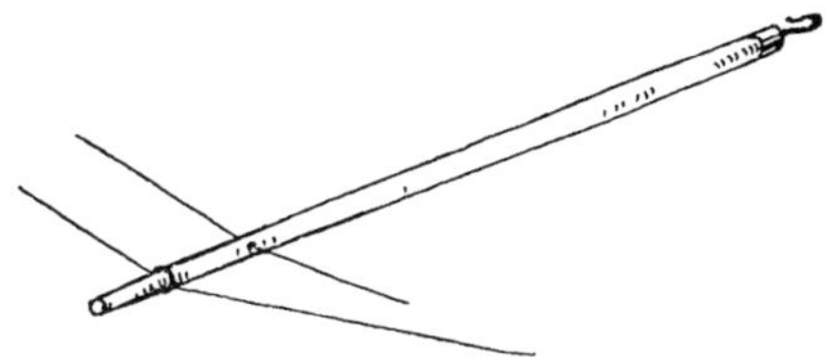

WHISKERS
Fig 94, Page 43

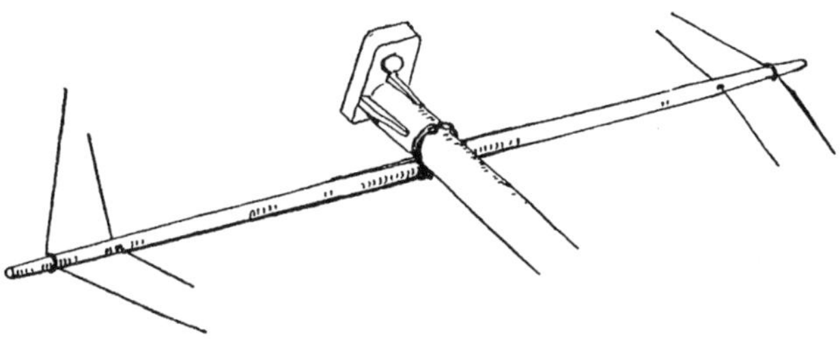

SPRITSAIL YARD
Fig 95, Page 43

WHISKERS
Fig. 96, Page 44

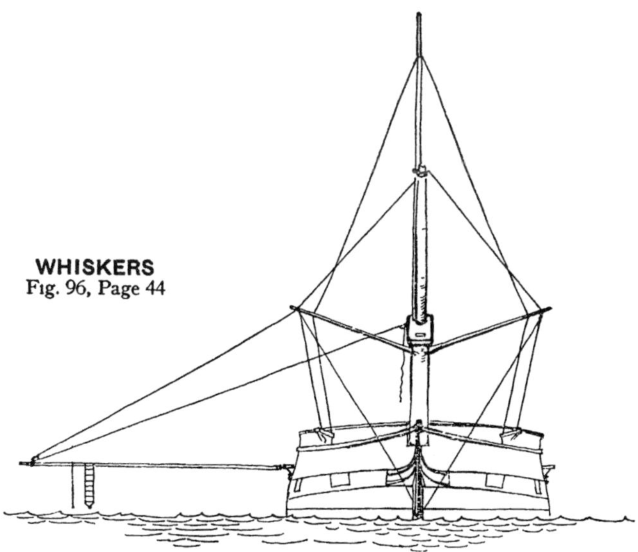

Plate 15

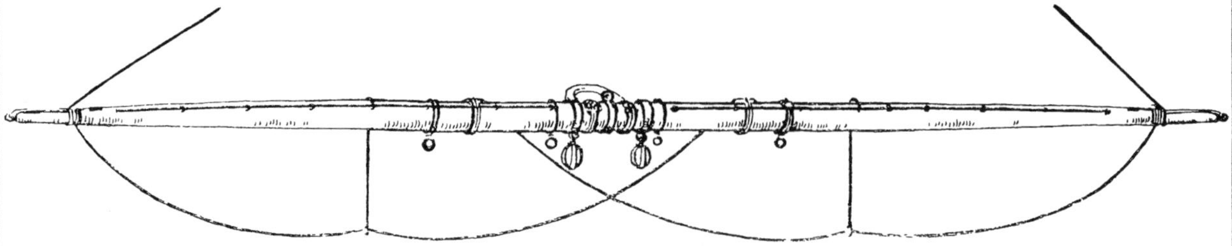

TOP-GALLANT YARD
Fig 97, Page 45

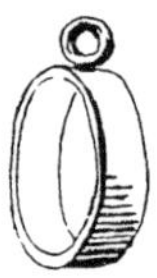

SLING BAND
Fig 98, Page 45

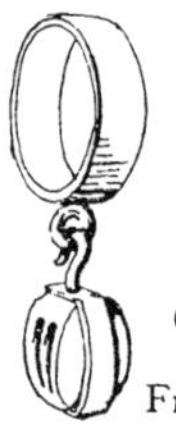

QUARTER BLOCK
Fig 99, Page 45

CLEW LINE STRAP
Fig 100, Page 45

STRAP FOR ROLLING ROPE
Fig 101, Page 46

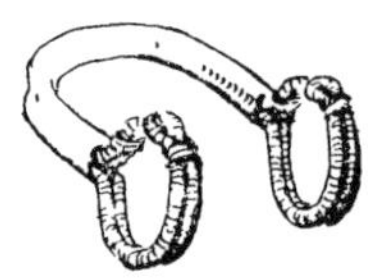

PARREL
Fig 102, Page 45

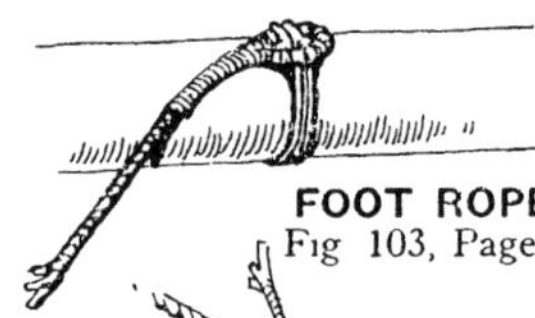

FOOT ROPES
Fig 103, Page 46

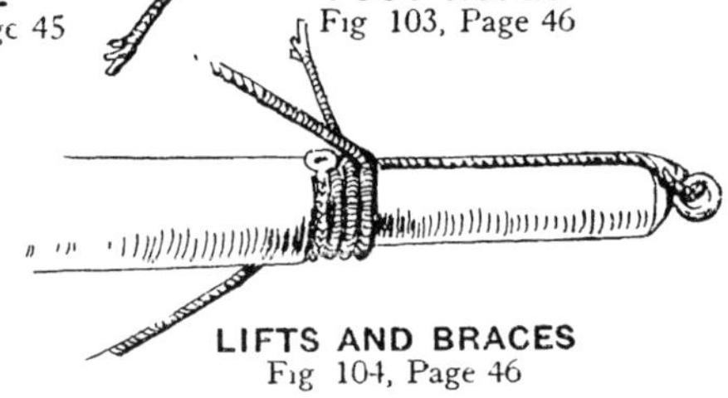

LIFTS AND BRACES
Fig 104, Page 46

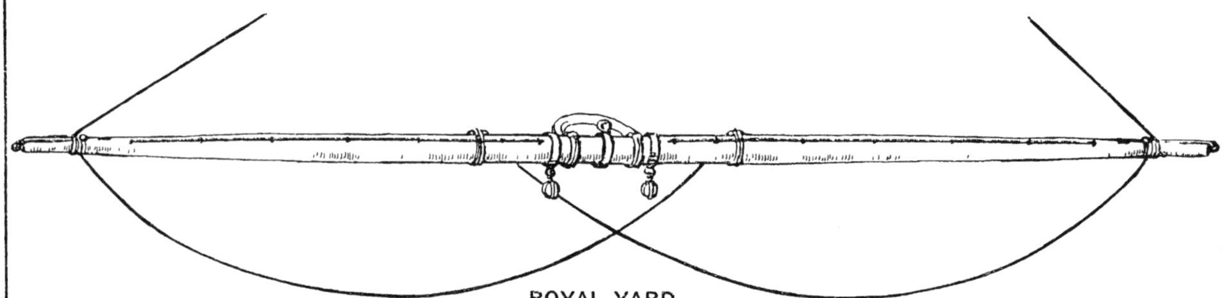

ROYAL YARD
Fig 105, Page 47

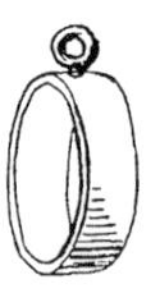

SLING BAND
Fig 106, Page 45

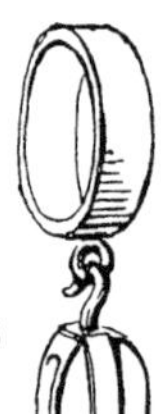

QUARTER BLOCK
Fig 107, Page 45

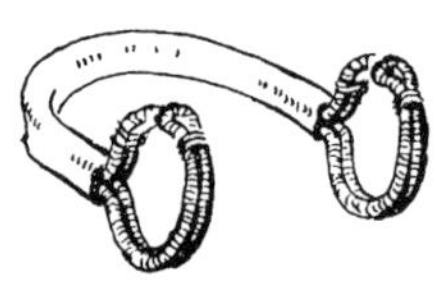

PARREL
Fig 108, Page 45

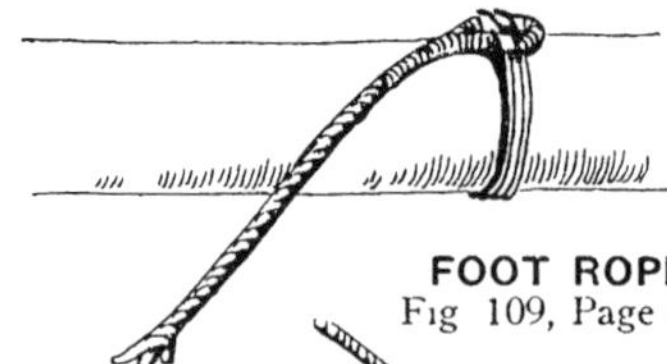

FOOT ROPE
Fig 109, Page 46

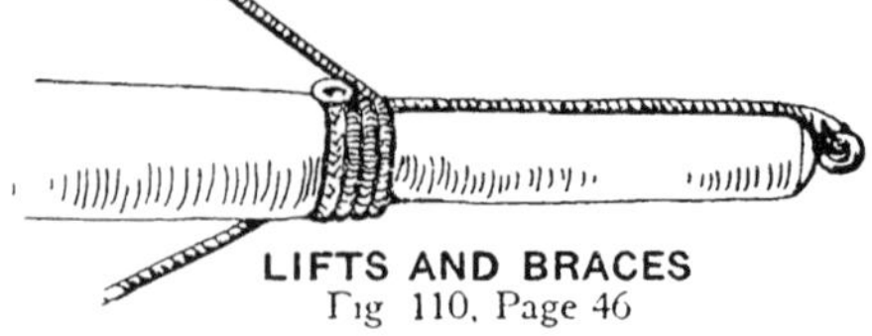

LIFTS AND BRACES
Fig 110, Page 46

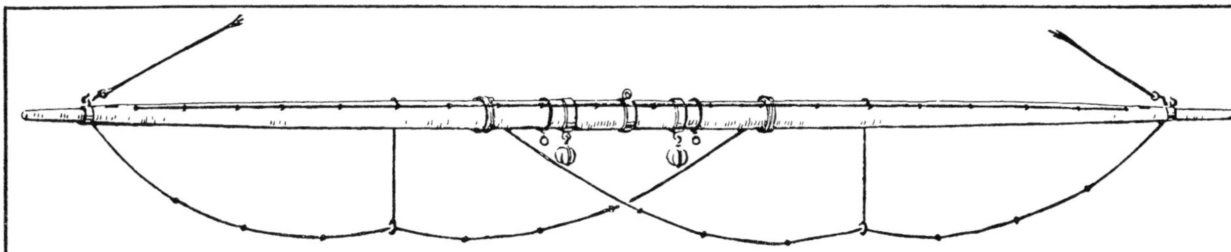

TOP-GALLANT YARD
Fig 111, Page 45

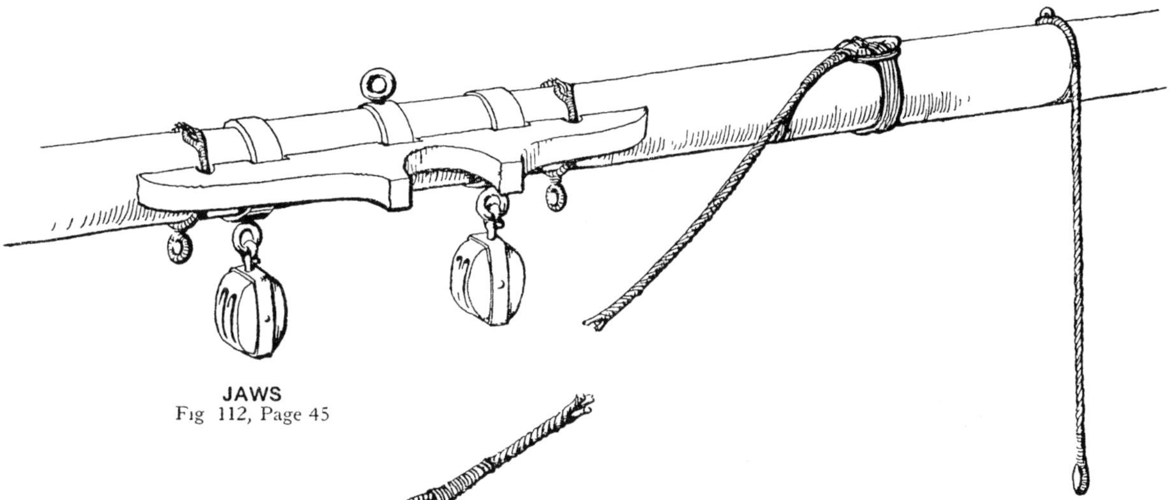

JAWS
Fig 112, Page 45

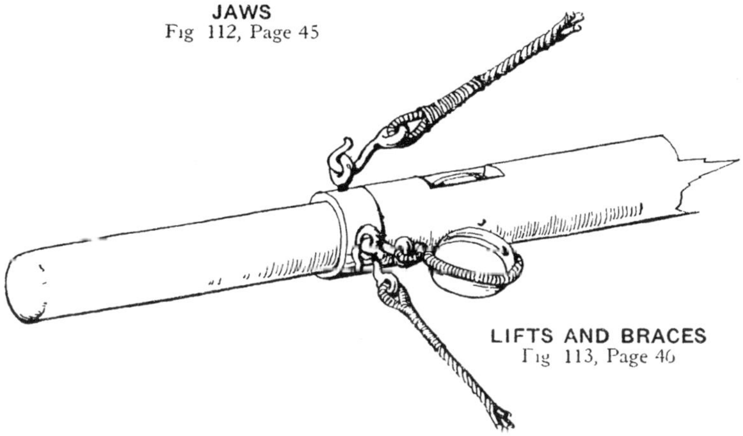

LIFTS AND BRACES
Fig 113, Page 46

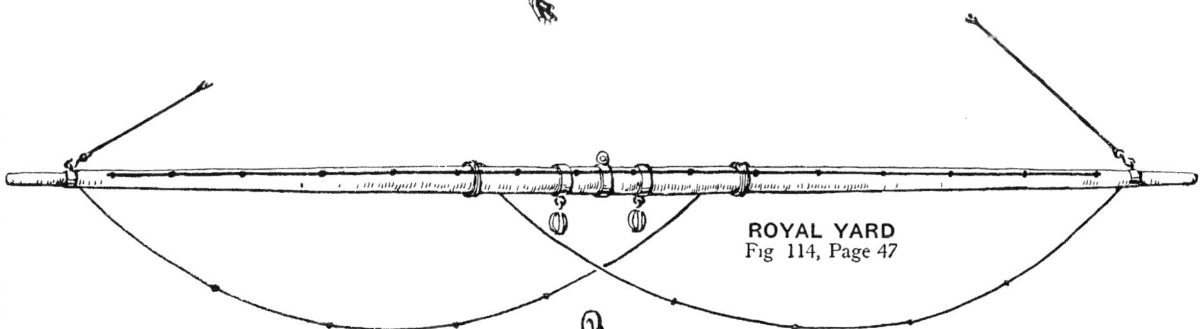

ROYAL YARD
Fig 114, Page 47

QUARTER BLOCK
Fig 115, Page 47

Plate 17

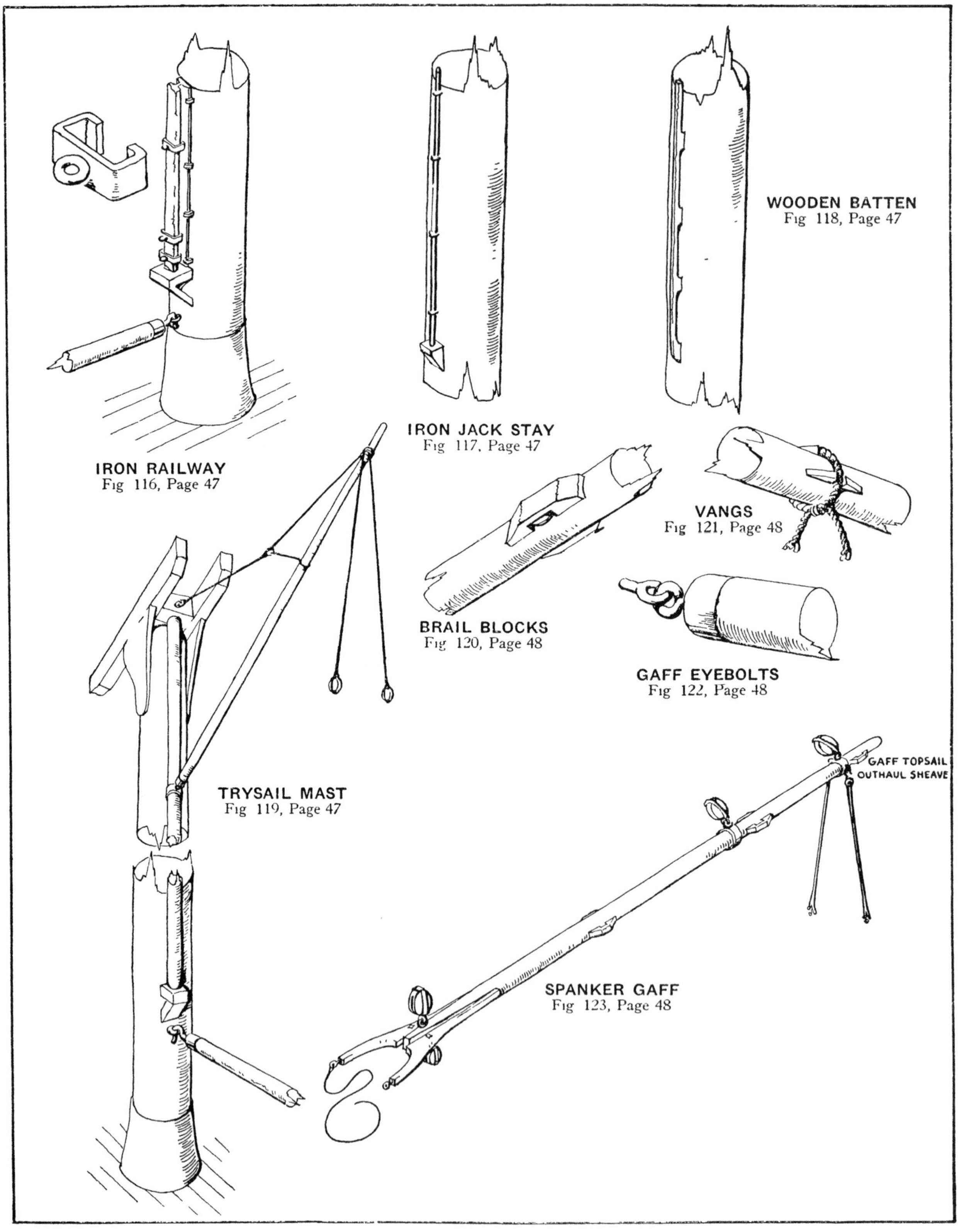
IRON RAILWAY
Fig 116, Page 47
IRON JACK STAY
Fig 117, Page 47
WOODEN BATTEN
Fig 118, Page 47
TRYSAIL MAST
Fig 119, Page 47
BRAIL BLOCKS
Fig 120, Page 48
VANGS
Fig 121, Page 48
GAFF EYEBOLTS
Fig 122, Page 48
SPANKER GAFF
Fig 123, Page 48
GAFF TOPSAIL OUTHAUL SHEAVE

Plate 18

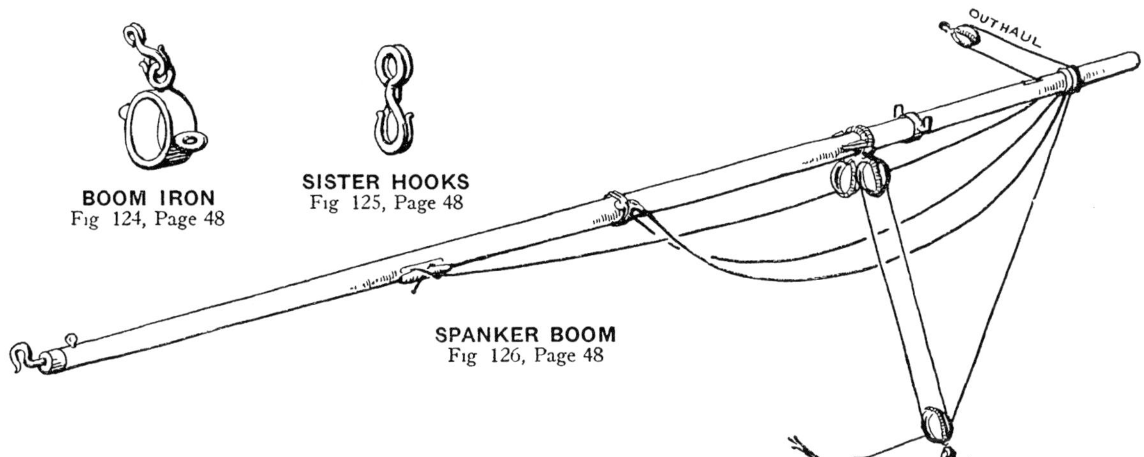

BOOM IRON
Fig 124, Page 48

SISTER HOOKS
Fig 125, Page 48

SPANKER BOOM
Fig 126, Page 48

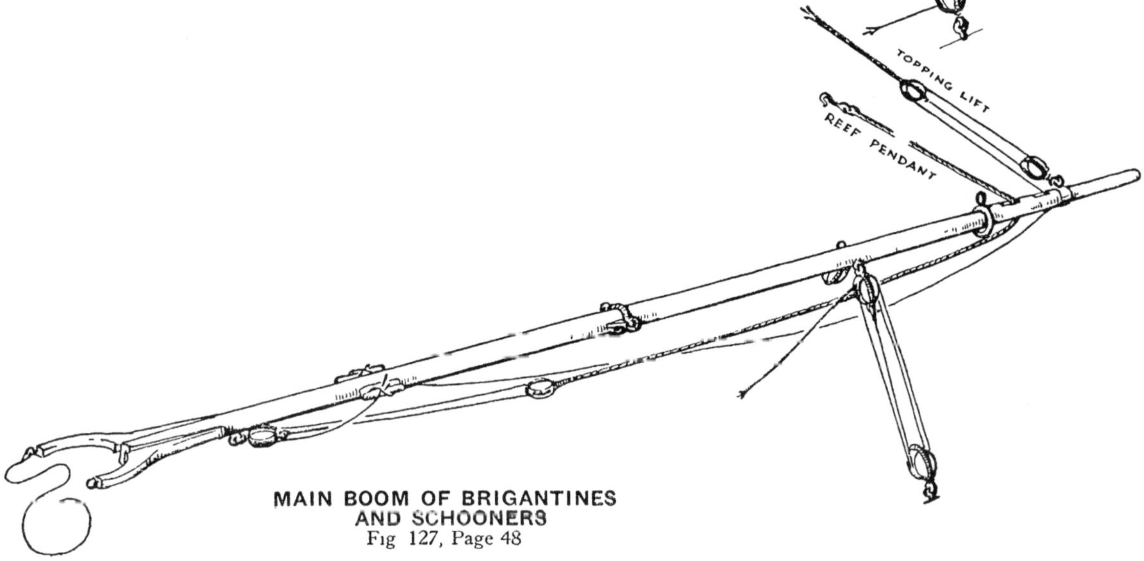

MAIN BOOM OF BRIGANTINES AND SCHOONERS
Fig 127, Page 48

TOP-GALLANT STUDDING-SAIL BOOM
Fig 128, Page 50

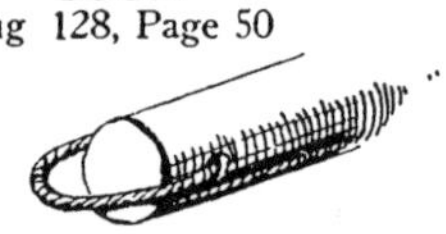

STUDDING-SAIL BOOM
Fig 129, Page 50

HALLIARD BLOCK
Fig 130, Page 50

TOPMAST STUDDING SAIL BOOM
Fig 131, Page 49

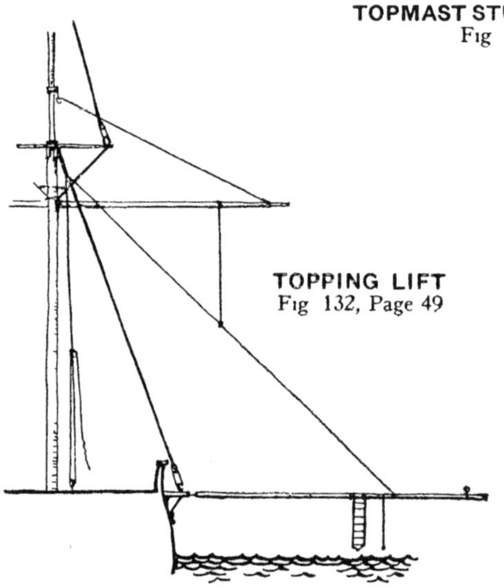

TOPPING LIFT
Fig 132, Page 49

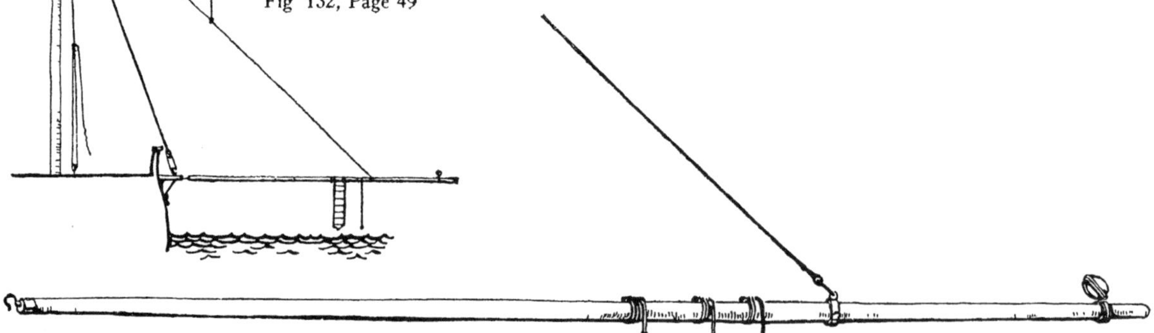

LOWER STUDDING SAIL BOOM
Fig 133, Page 49

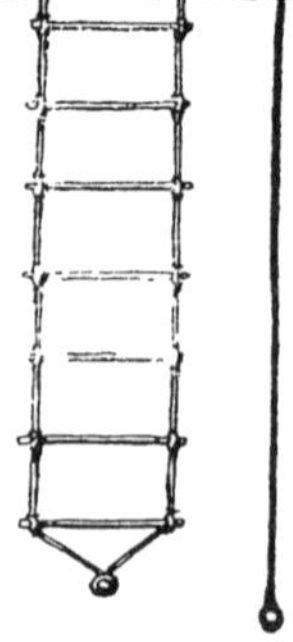

MOORING PENDANT
Fig 134, Page 49

IRON BAND
Fig 135, Page 49

OUT HAUL BLOCK
Fig 136, Page 49

Plate 20

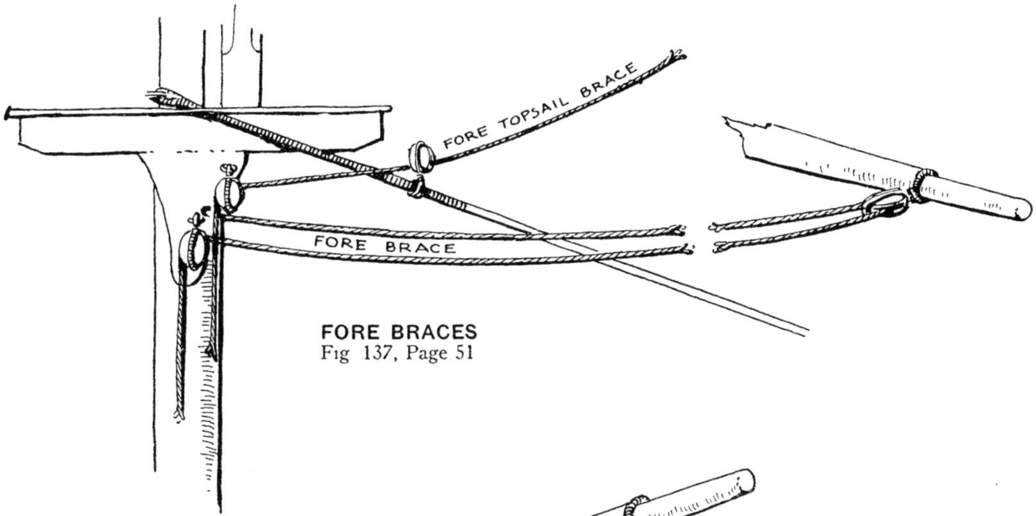

FORE BRACES
Fig 137, Page 51

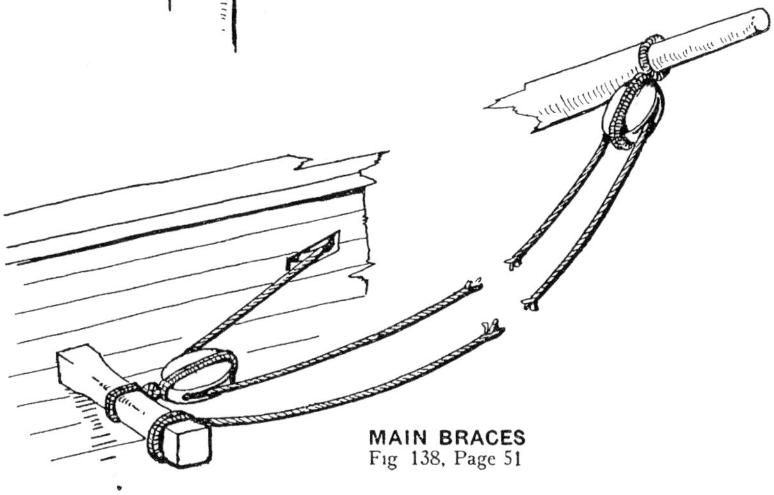

MAIN BRACES
Fig 138, Page 51

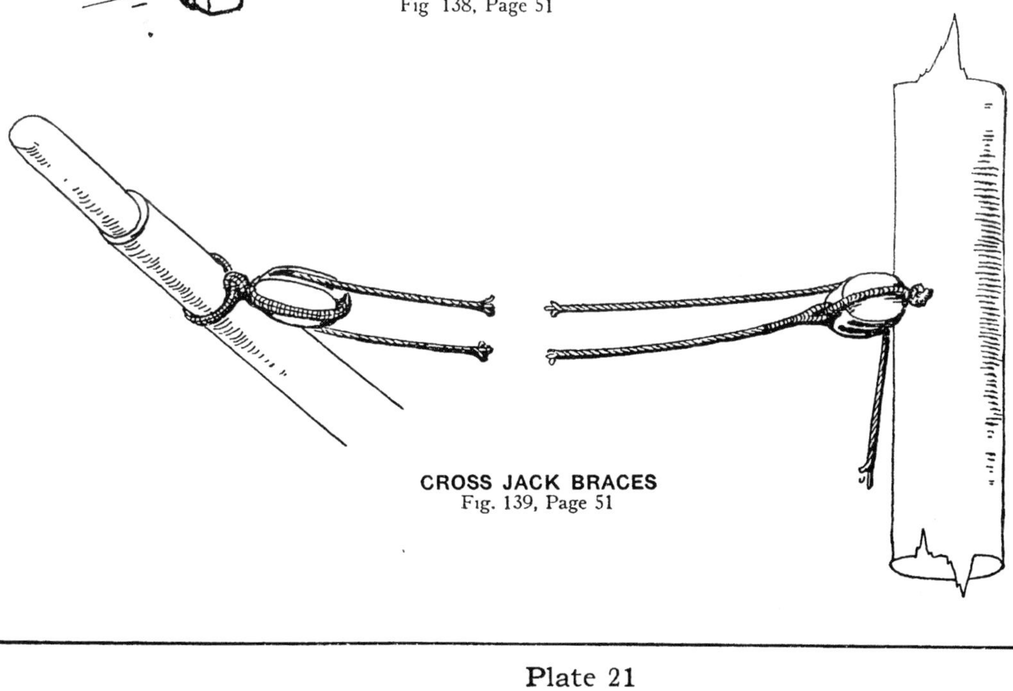

CROSS JACK BRACES
Fig. 139, Page 51

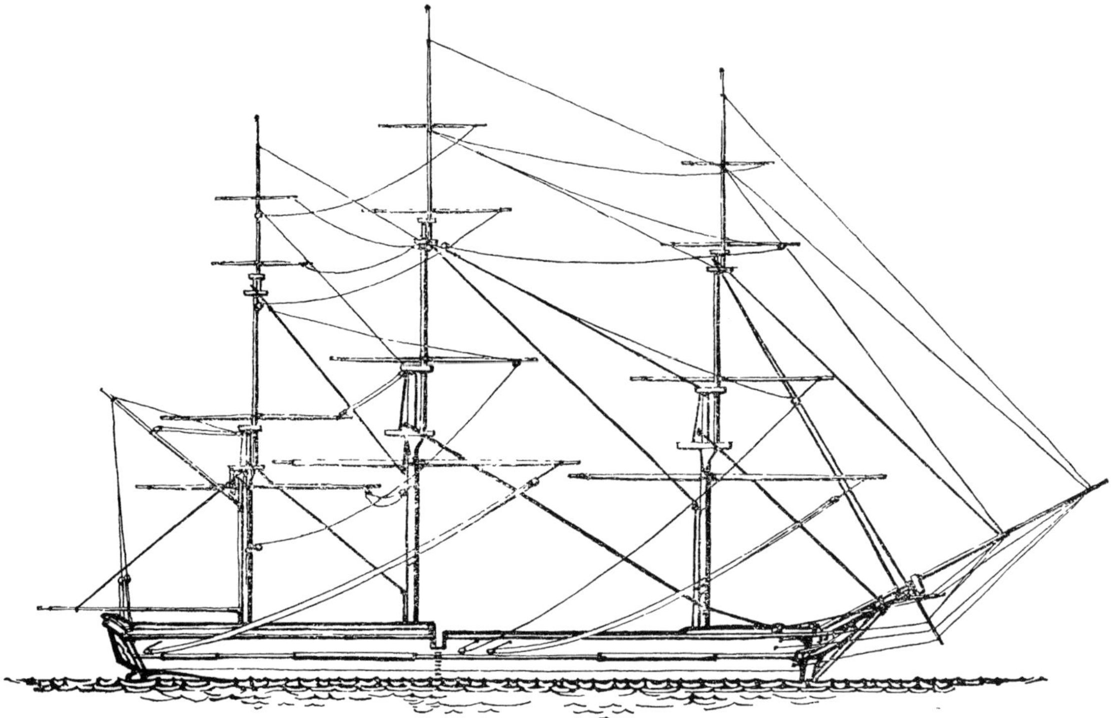

BRACES
Fig 145, Pages 51-52-53

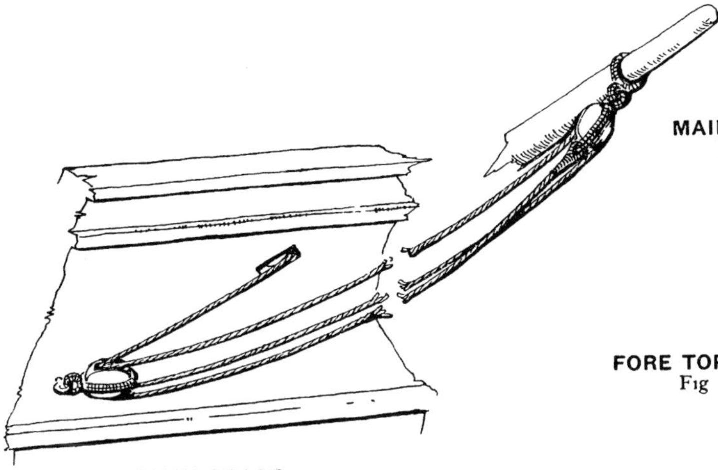

MAIN BRACE
Fig 141, Page 51

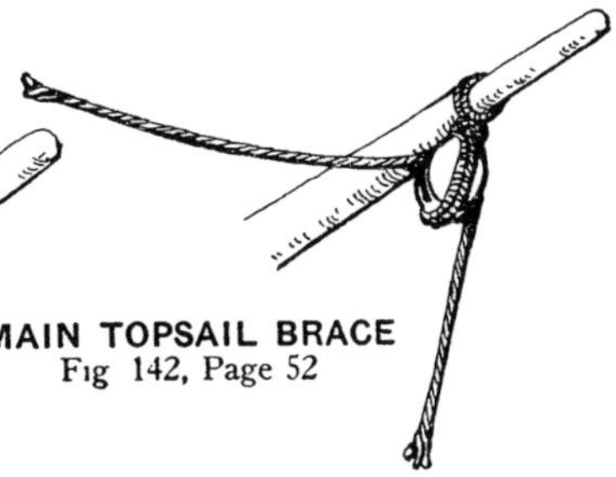

MAIN TOPSAIL BRACE
Fig 142, Page 52

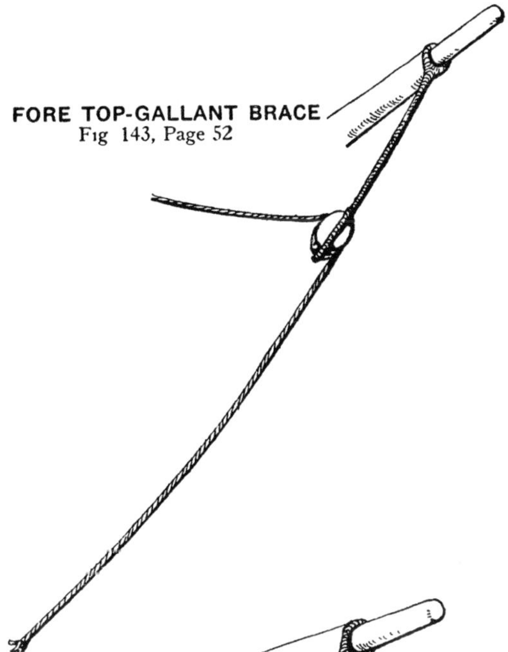

FORE TOP-GALLANT BRACE
Fig 143, Page 52

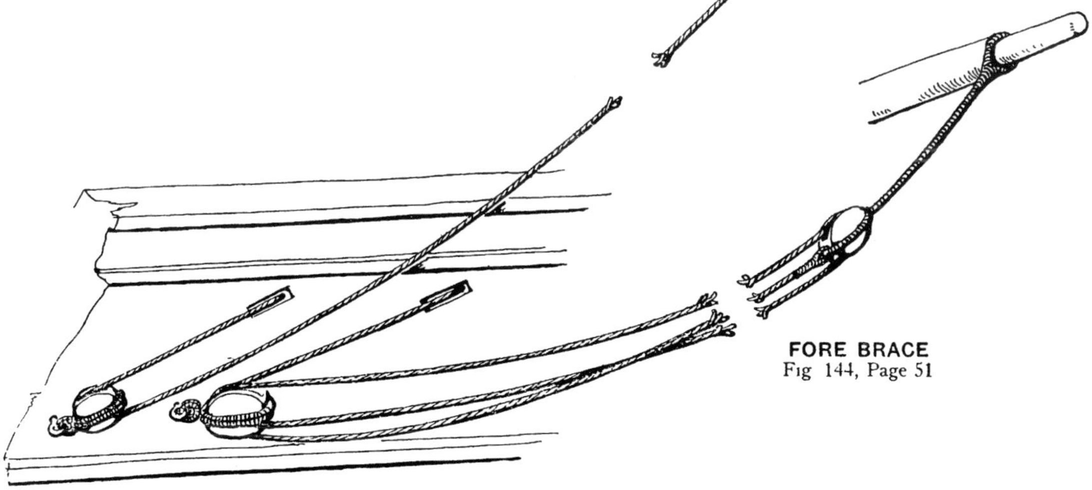

FORE BRACE
Fig 144, Page 51

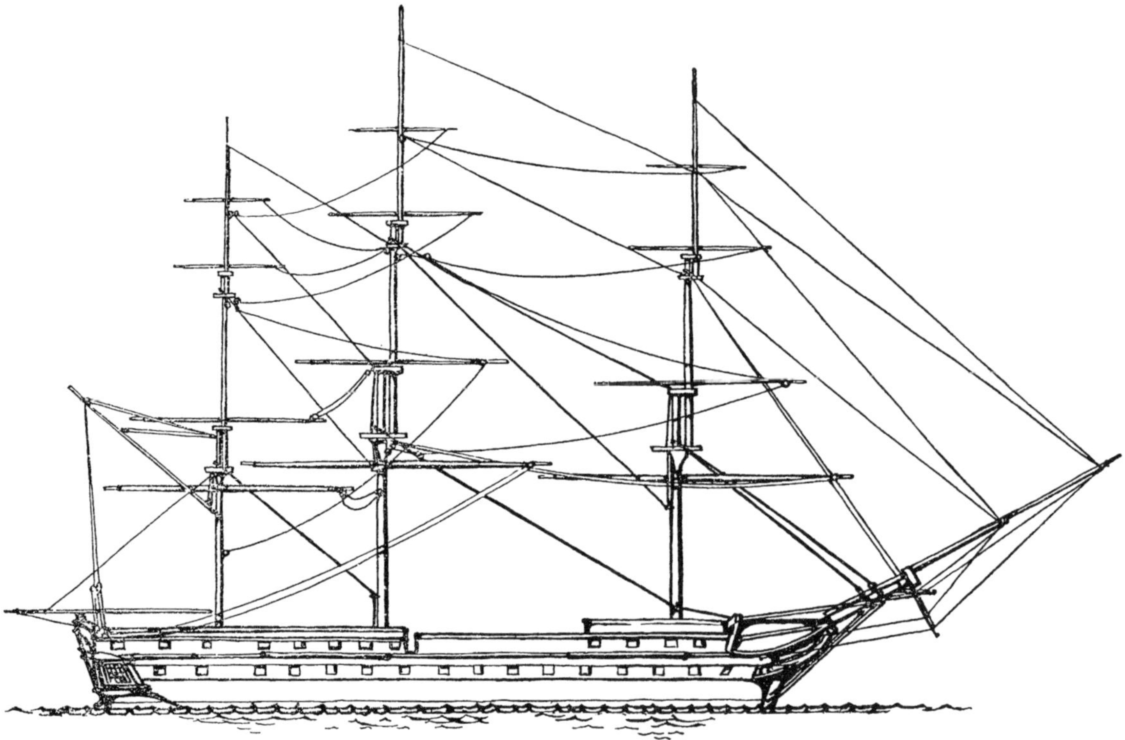

BRACES
Fig 140, Pages 51-52-53

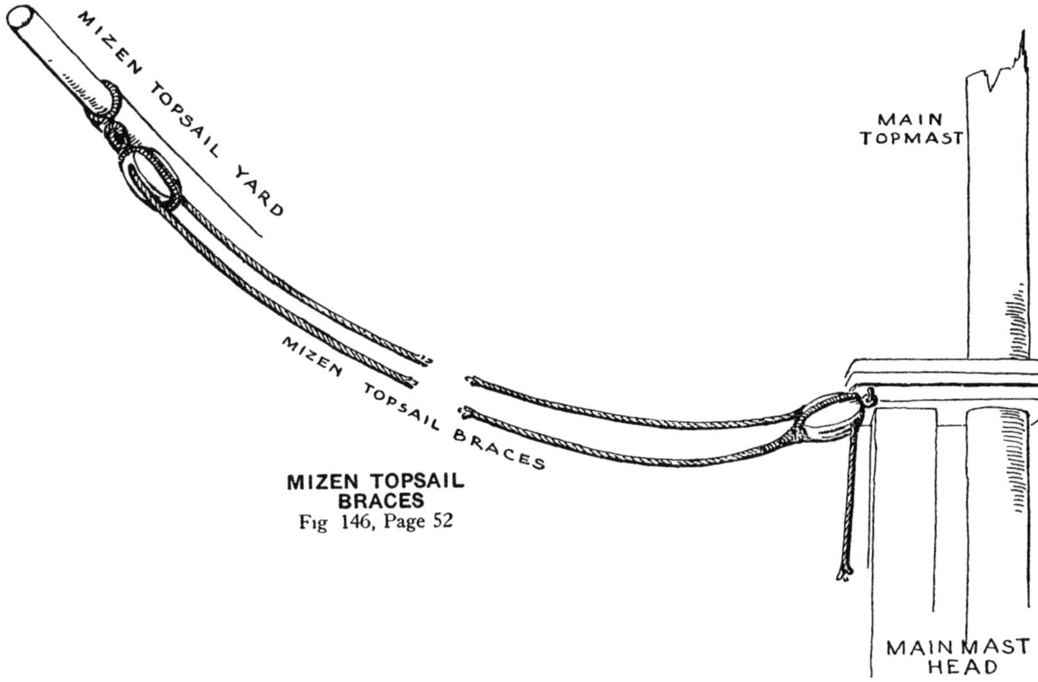

MIZEN TOPSAIL BRACES
Fig 146, Page 52

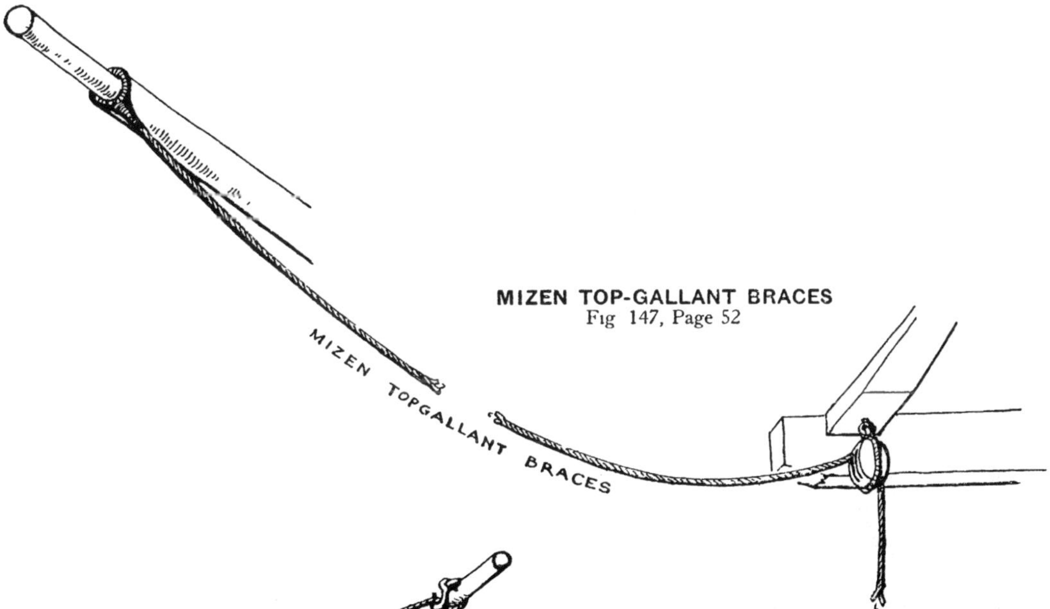

MIZEN TOP-GALLANT BRACES
Fig 147, Page 52

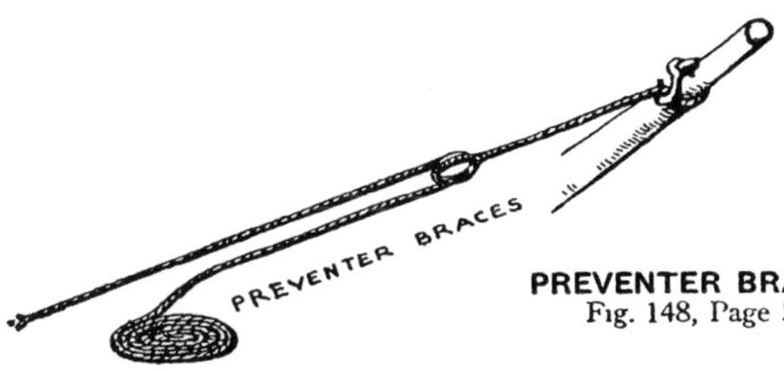

PREVENTER BRACES
Fig. 148, Page 53

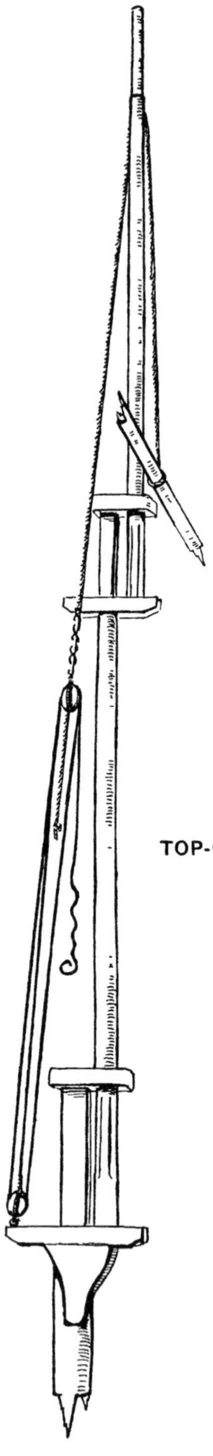

TOP-GALLANT HALLIARDS
Fig 149, Page 53

ROYAL HALLIARDS
Fig 150, Page 53

Plate 26

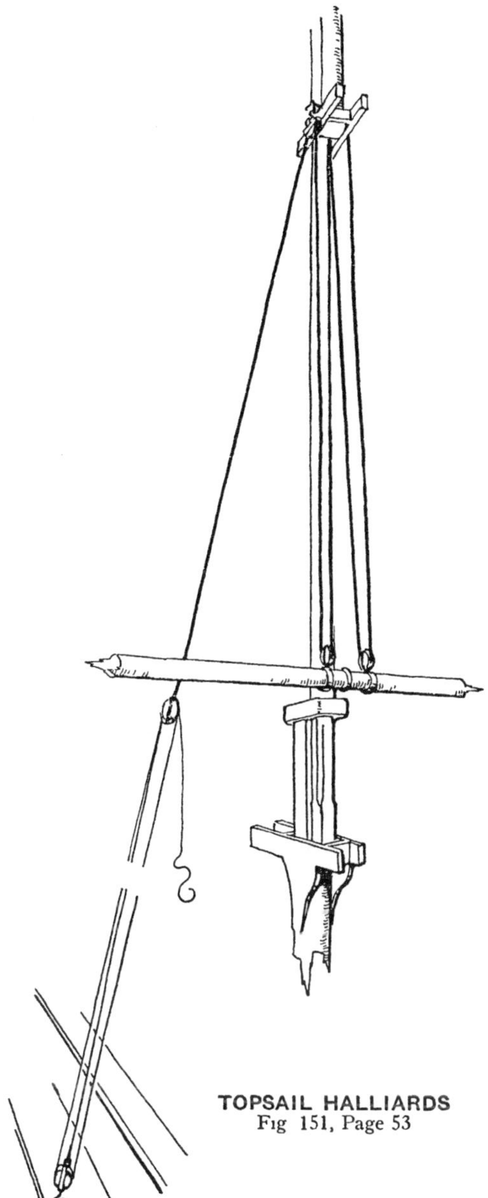

TOPSAIL HALLIARDS
Fig 151, Page 53

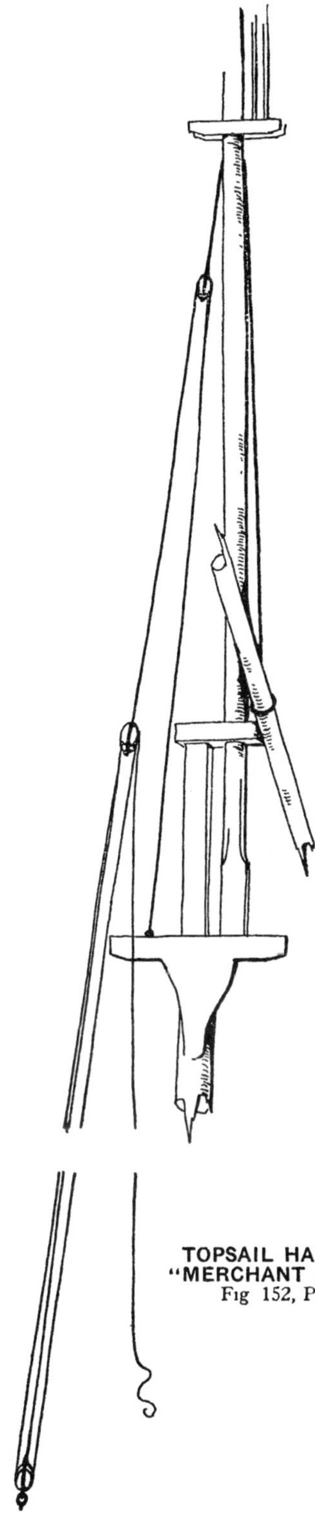

TOPSAIL HALLIARDS "MERCHANT SERVICE"
Fig 152, Page 53

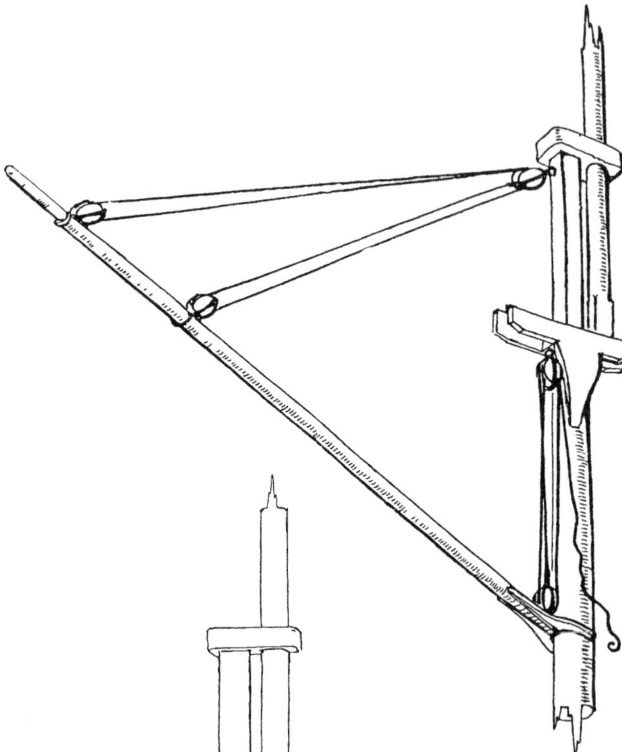

THROAT AND PEAK HALLIARDS
Fig 153, Page 53- 54

STORM STAYSAIL HALLIARD
Fig 154, Page 54

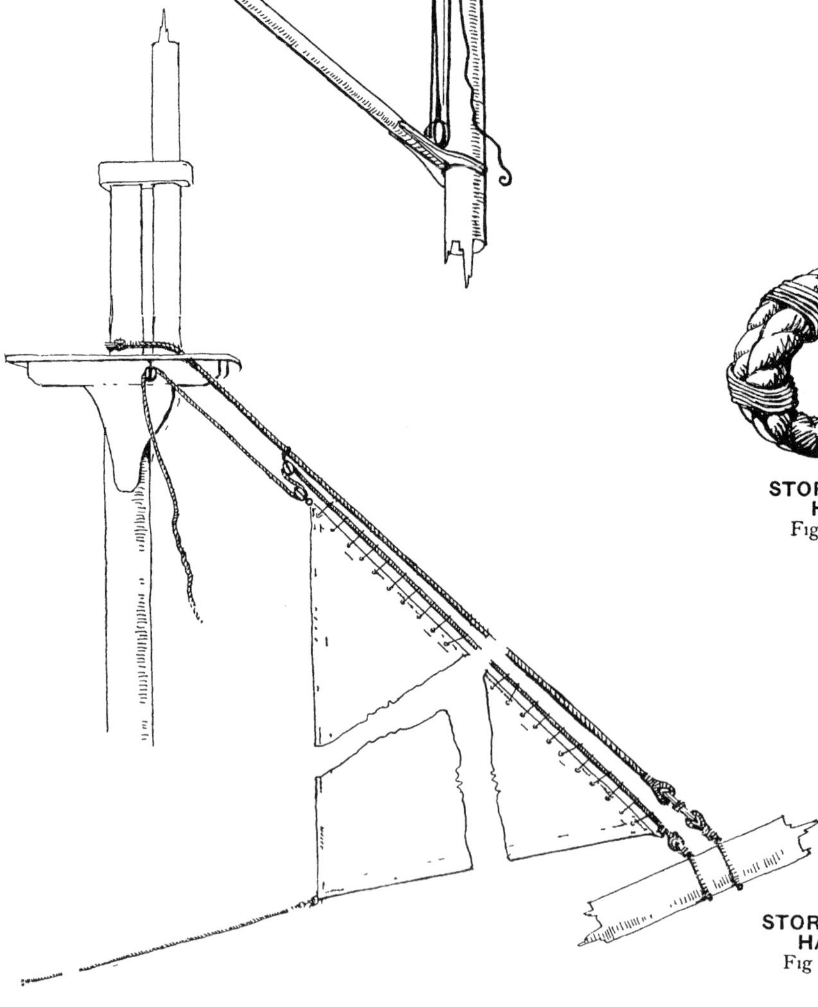

STORM STAYSAIL HALLIARDS
Fig 155, Page 54

Plate 28

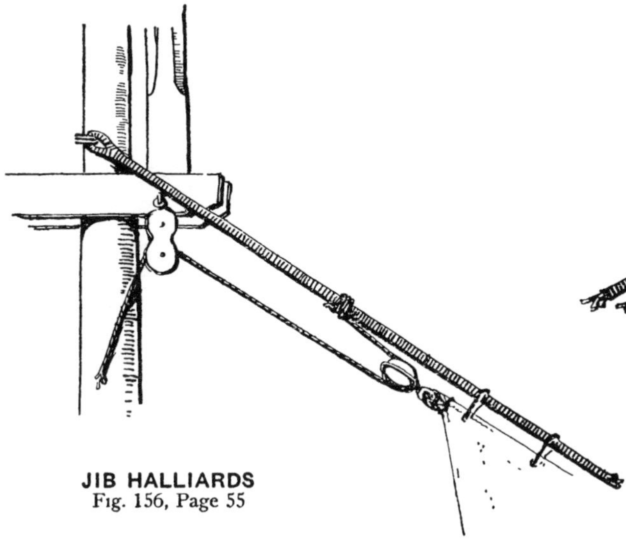

JIB HALLIARDS
Fig. 156, Page 55

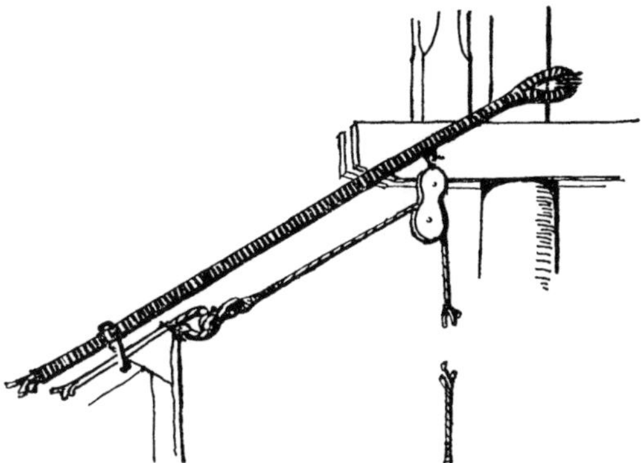

JIB AND TOPMAST STAYSAIL HALLIARDS
Fig 157, Page 55

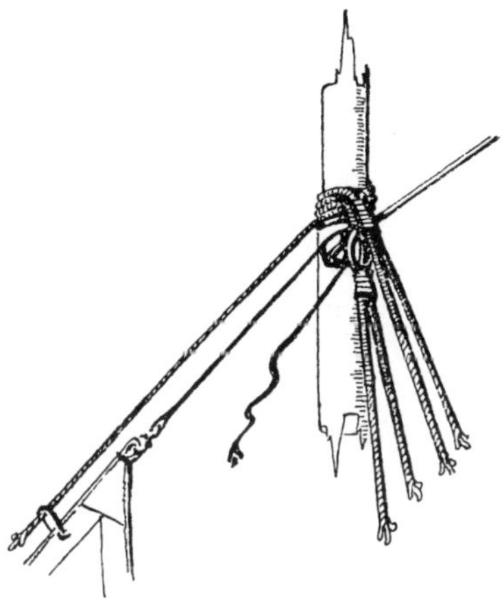

FLYING JIB HALLIARDS
Fig 158, Page 55

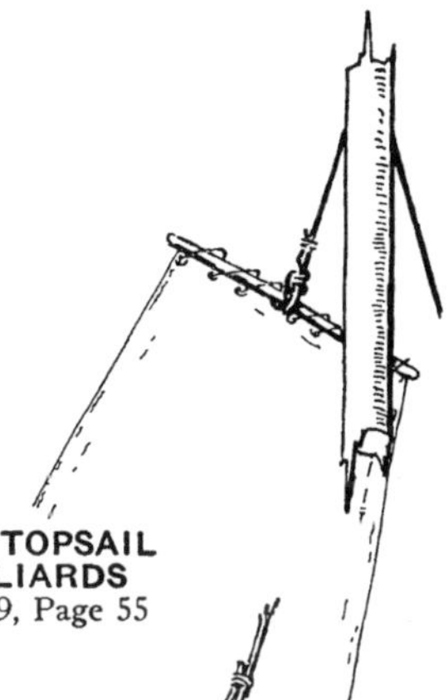

GAFF TOPSAIL HALLIARDS
Fig 159, Page 55

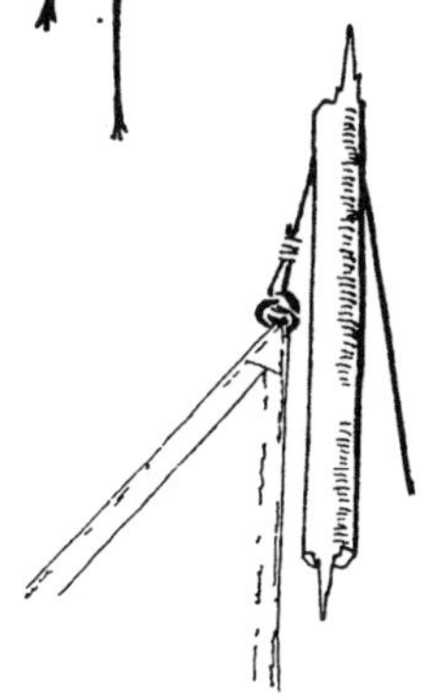

GAFF TOPSAIL HALLIARDS
Fig 161, Page 55

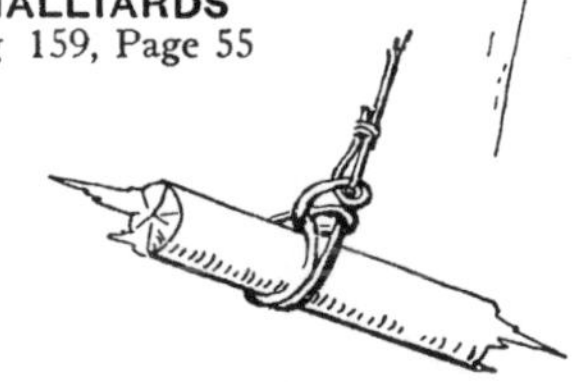

FISHERMAN'S BEND GAFF TOPSAIL HALLIARDS
Fig 160, Page 55

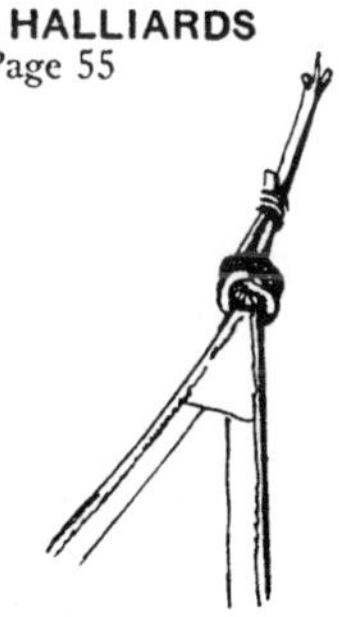

SHEET BEND GAFF TOPSAIL HALLIARDS
Fig 162, Page 55

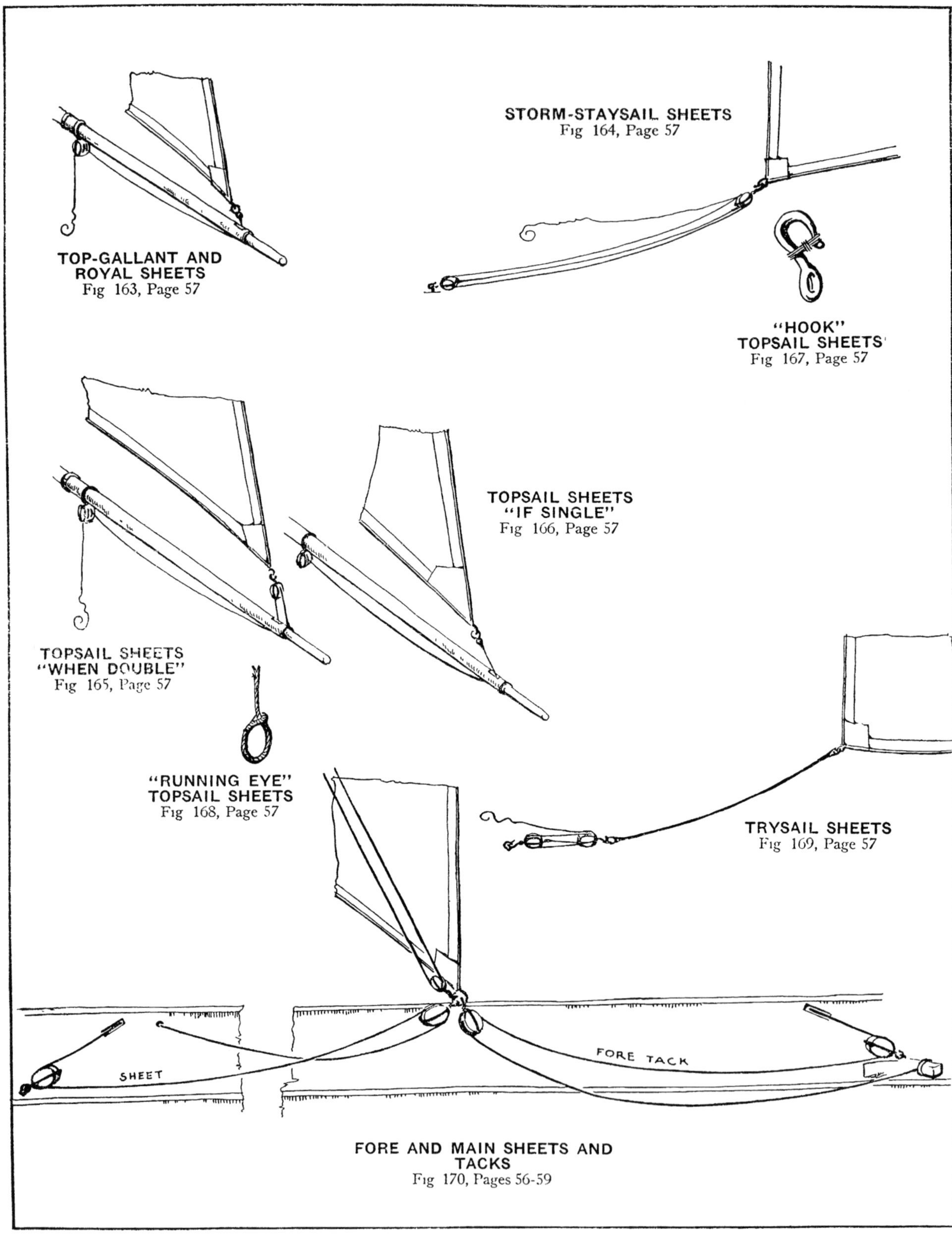

TOP-GALLANT AND ROYAL SHEETS
Fig 163, Page 57

STORM-STAYSAIL SHEETS
Fig 164, Page 57

"HOOK" TOPSAIL SHEETS
Fig 167, Page 57

TOPSAIL SHEETS "WHEN DOUBLE"
Fig 165, Page 57

TOPSAIL SHEETS "IF SINGLE"
Fig 166, Page 57

"RUNNING EYE" TOPSAIL SHEETS
Fig 168, Page 57

TRYSAIL SHEETS
Fig 169, Page 57

FORE AND MAIN SHEETS AND TACKS
Fig 170, Pages 56-59

Plate 30

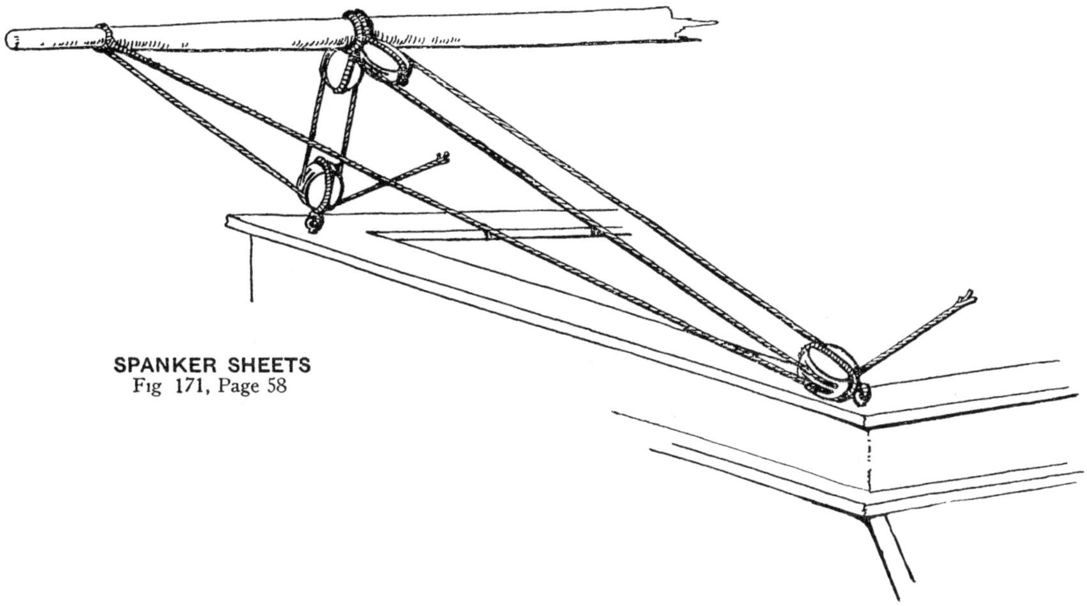

SPANKER SHEETS
Fig 171, Page 58

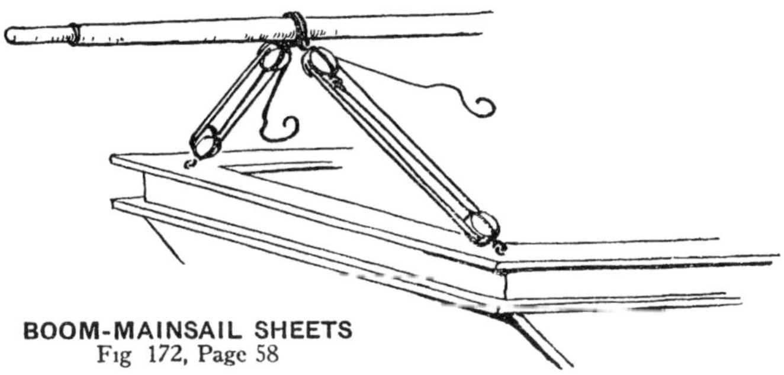

BOOM-MAINSAIL SHEETS
Fig 172, Page 58

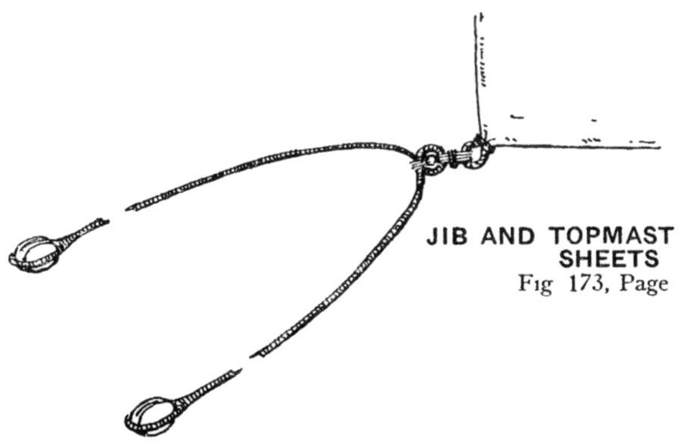

JIB AND TOPMAST STAYSAIL SHEETS
Fig 173, Page 57

GAFF TOPSAIL SHEETS
Fig 174, Page 58

Plate 31

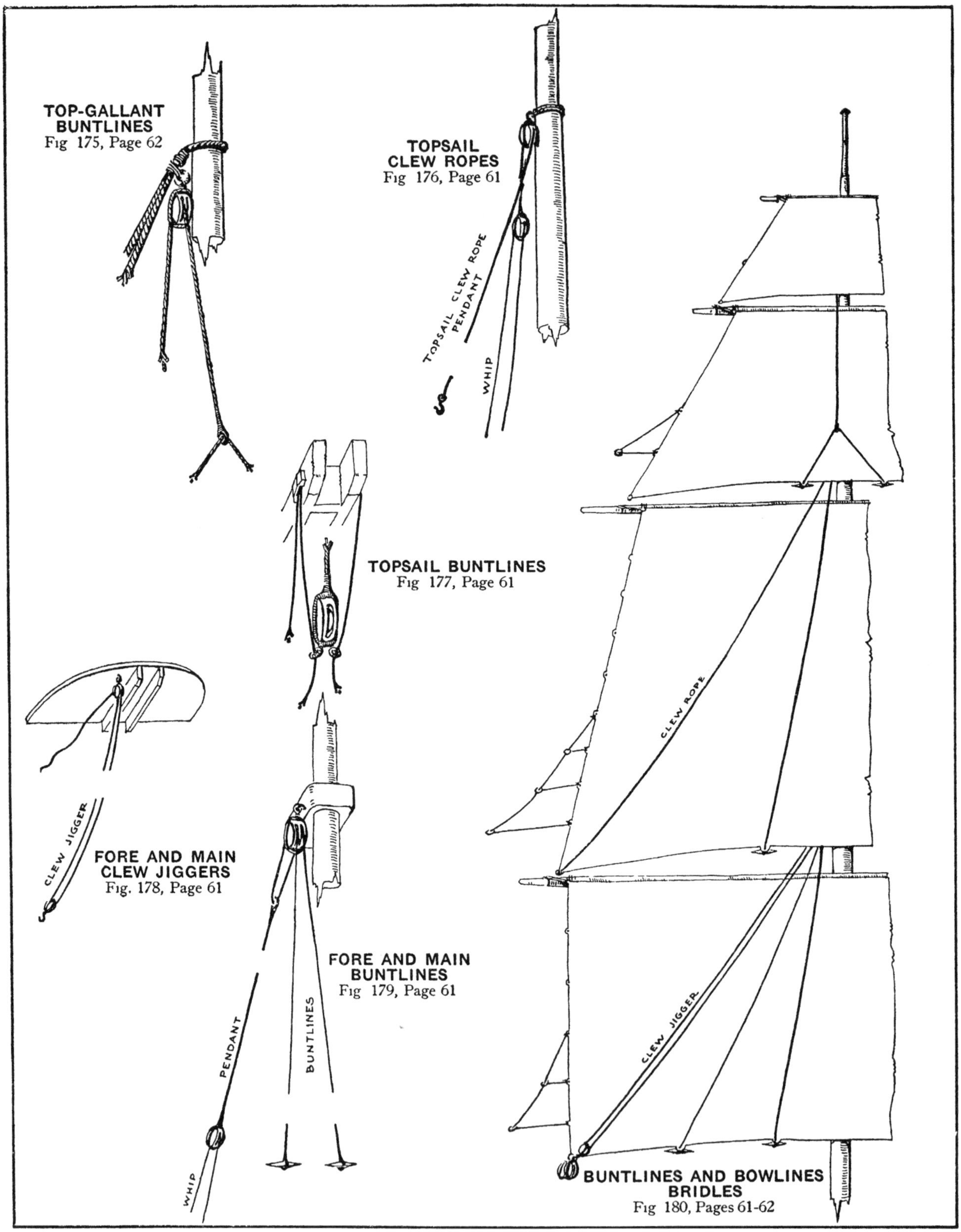

TOP-GALLANT BUNTLINES
Fig 175, Page 62

TOPSAIL CLEW ROPES
Fig 176, Page 61

TOPSAIL BUNTLINES
Fig 177, Page 61

FORE AND MAIN CLEW JIGGERS
Fig. 178, Page 61

FORE AND MAIN BUNTLINES
Fig 179, Page 61

BUNTLINES AND BOWLINES BRIDLES
Fig 180, Pages 61-62

Plate 32

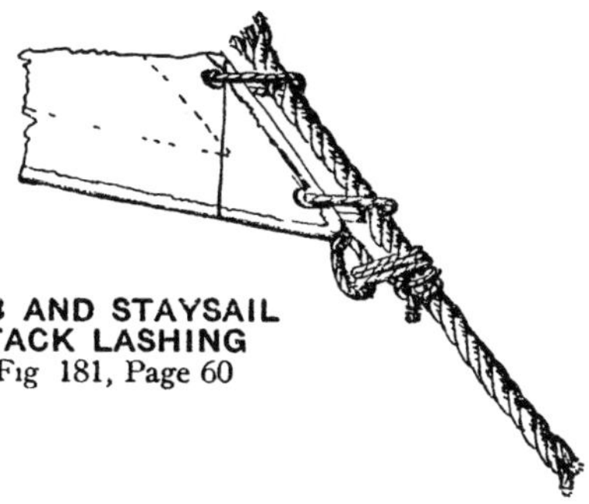

JIB AND STAYSAIL TACK LASHING
Fig 181, Page 60

CLEW GARNETS SECRET BLOCKS
Fig 182, Page 60

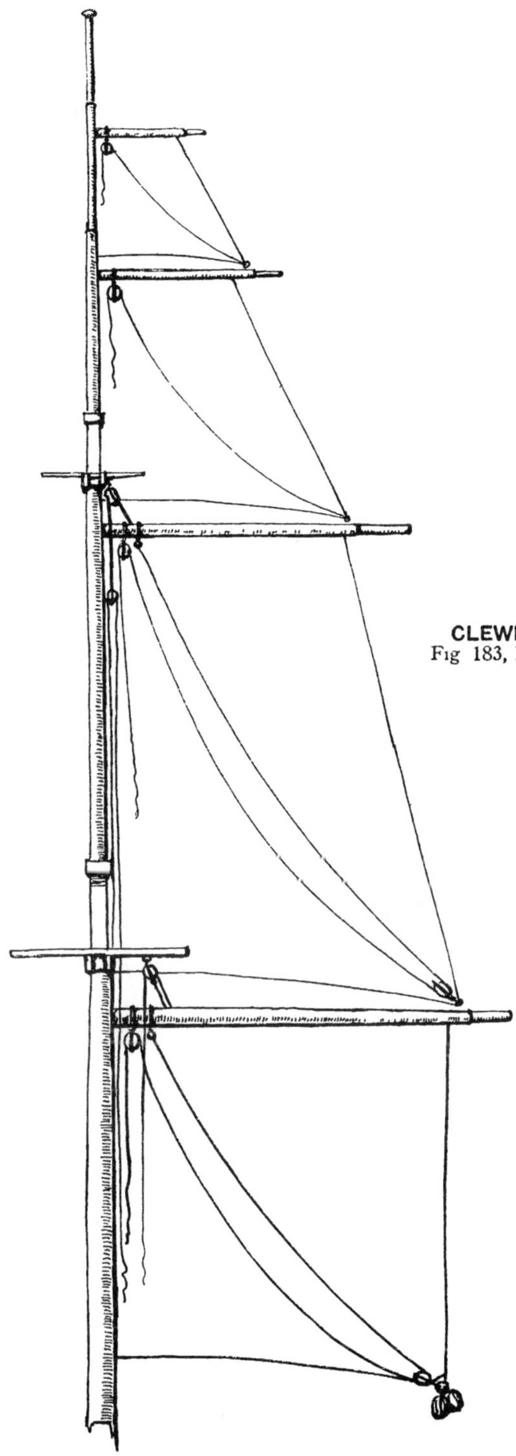

CLEWLINE
Fig 183, Page 60

BOWLINES
Fig 184, Page 62

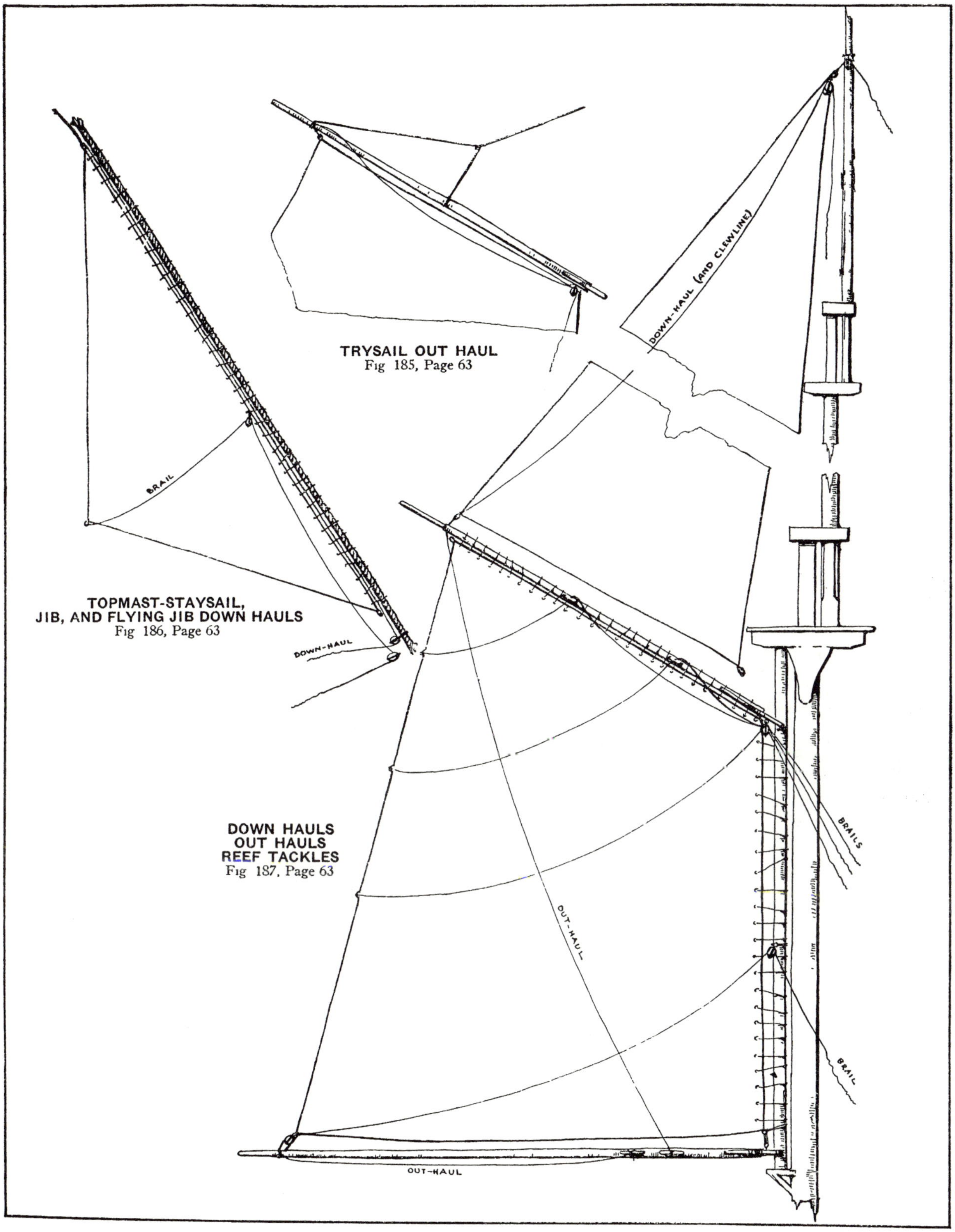

TRYSAIL OUT HAUL
Fig 185, Page 63

TOPMAST-STAYSAIL, JIB, AND FLYING JIB DOWN HAULS
Fig 186, Page 63

DOWN HAULS OUT HAULS REEF TACKLES
Fig 187, Page 63

Plate 35

REEF TACKLES AND LEECH LINES
Fig 188, Pages 63-64

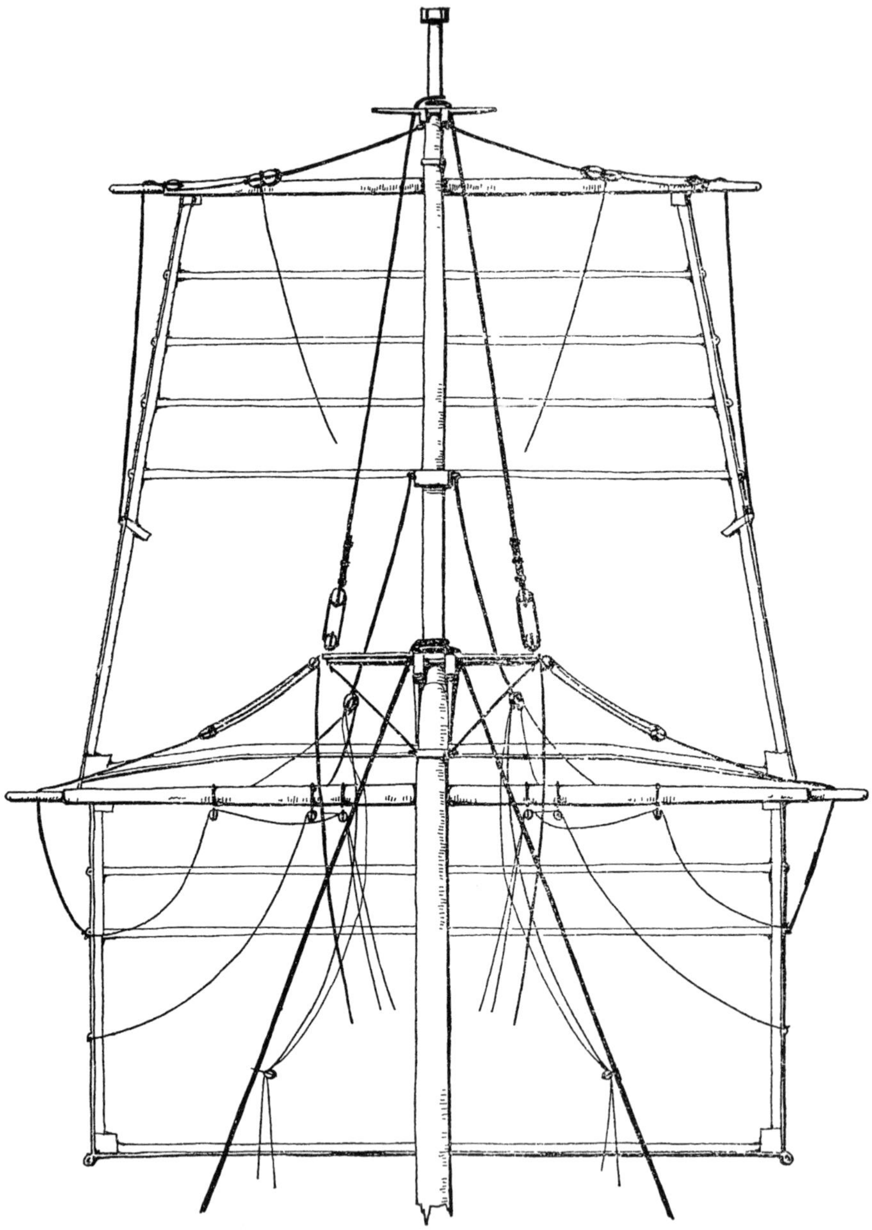

Plate 36

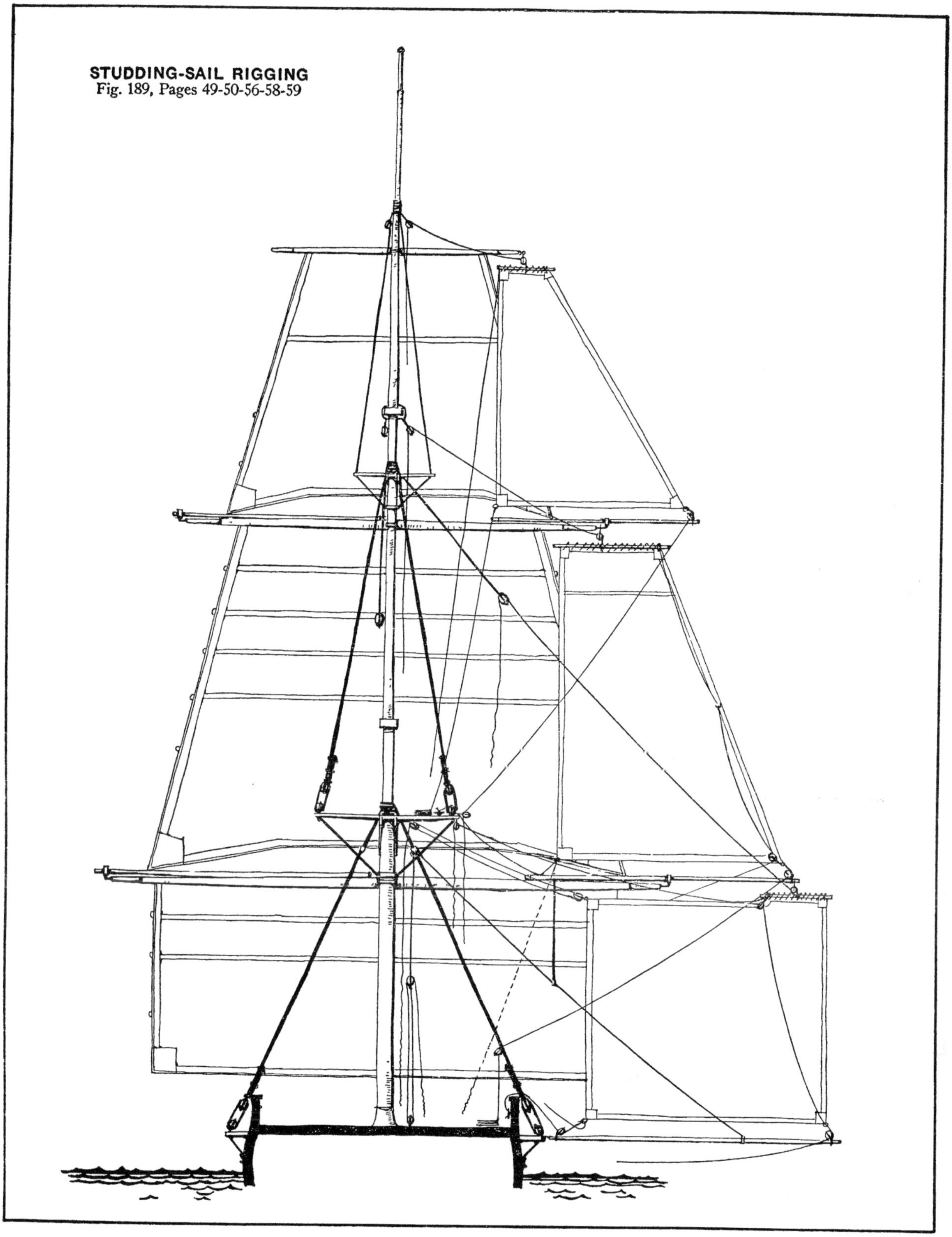

STUDDING-SAIL RIGGING
Fig. 189, Pages 49-50-56-58-59

Plate 37